14th International Conference on Advances in Mass Data Analysis of Images and Signals in Medicine, Biotechnology, Chemistry and Food Industry (MDA 2019)

New York, New York, USA
13-16 July 2019

Editors:

Petra Perner

ISBN: 978-1-5108-9521-8

Printed from e-media with permission by:

Curran Associates, Inc.
57 Morehouse Lane
Red Hook, NY 12571

ibai-publishing
Arno-Nitzsche-Str. 45
04277 Leipzig, Germany

info@ibai-publishing.org

Additional copies of this publication are available from:

Curran Associates, Inc.
57 Morehouse Lane
Red Hook, NY 12571 USA
Phone: 845-758-0400
Fax: 845-758-2633
Email: curran@proceedings.com
Web: www.proceedings.com

Table of Content

Telemedicine Data Flow Models
Calin Ciufudean *1*

Image analysis technique and procedure for its application in 2D
Javier Bilbao, Imanol Bilbao, Andoni Olozaga and Cristina Feniser *10*

Indirect Determination of Mycotoxin Concentration with Image Processing and Data Mining
Petra Perner *23*

The Improved Algorithm of Barrier Tree in RNA Folding Structure Based on Basin Hopping Graph
Zhendong Liu, Gang Li, and Patrick Wang *31*

Automatic Liver Segmentation Using shape context constraint Network
Lifang Zhou, Lu Wang, Patrick Wang, Weisheng Li, Bangjun Lei ,
and Xueyuan Deng *41*

Author´s Index *49*

Petra Perner (Ed.)

Advances in Mass Data Analysis of Images and Signals

with Applications in Medicine, r/g/b Biotechnology, Food Industries and Dietetics, Biometry and Security, Agriculture, Drug Discover, and System Biology

14th International Conference, MDA 2019
New York, USA, July 13-16, 2019

Proceedings

ibai Publishing
www.ibai-publishing.org

Petra Perner (Ed.)

Advances in Mass Data Analysis of Images and Signals

with Applications in Medicine, r/g/b Biotechnology, Food Industries and Dietetics, Biometry and Security, Agriculture, Drug Discover, and System Biology Medicine, Biotechnology, Chemistry and Food Industry

14th International Conference, MDA 2019
New York, USA, July 13-16, 2019

Proceedings

ibai - publishing

Preface

The 14th International Conference on Mass Data Analysis of Signals and Images with Applications in Medicine, r/g/b Biotechnology, Food Industries and Dietetics, Biometry and Security, and Agriculture MDA 2019 held in July in Newark USA showed once more that the event is a must for all specialists from research and industry alike who like to stay informed about hot new topics in mass data analysis of signals and images. MDA 2019 had two invited talks, regular sessions, and invited sessions on specific very important topics. Altogether sixteen talks have been given.

The topics range of MDA from Case-Based Reasoning and Data Mining for 1-,2-,3-Signals, Time Series Analysis, Aspects of Mass Data Analysis, Telemedicine, and Spectrometer Signal Analysis. Full research work and work in progress has been presented.

The first invited talk was on "Fuzzy Recurrence Analysis of Complex Systems Dynamics" given by Prof. Dr. Tuan D. Pham from the Department of Biomedical Engineering of the Linkoping University in Sweden. The second invited talk was on "Intelligent Pattern Recognition (IPR) and Applications" given by Prof. Dr. Patrick S.P. Wang, Fellow IAPR, ISIBM, WASE&IETI and IEEE ISIBM outstanding achievement awardee from the Northeastern University in Boston, USA.

The discussion made the conference very inspiring and influencing. A lot of ideas were exchanged that will help to progress in further work.

The social program gave a great opportunity to network, interact among each other, and exchange ideas about certain research aspects. All together it was a very inspiring and successful conference.

Five full papers have been selected for publishing in the proceedings "Advances in Advances in Mass Data Analysis of Images and Signals" (ISBN 978-3-942952-64-4) by ibai-publishing house (ibai-publishing.org).

Selected revised papers of this conference as well as contributed papers will be published in the October issue of the International Journal Transaction on Mass Data Analysis of Signals and Images (www.ibai-publishing.org).

We are happy to see that new automatic systems for the analysis of signals and images have been presented at MDA and that a lot of researchers followed our mission to bring new signal and image analysis methods into real applications that introduce a new quality level to various real life applications. We decided to bring aware MDA to a broader audience. Therefore, we changed the title of MDA to Mass Data Analysis of Signals and Images in Artificial Intelligence and Pattern Recognition MDA-AI&PR that should attract more people to the mission of MDA.

The 15th conference will be held in 2020 in New York (www.mda-signals.de) under the auspices of the World Congress Frontiers on Intelligent Data Analysis DSA 2020 (www.worldcongressdsa.com) and it is now named MDA-AI&PR 2020.

We would like to invite you to contribute to this conference. Please come and join us.

July, 2019

Prof. Dr. Petra Perner
Chair of the Conference MDA

The German National Library listed this publication in the German National Bibliography.
Detailed bibliographical data can be downloaded from http://dnb.ddb.de.

ibai-publishing
Prof. Dr. Petra Perner
PF 30 11 38
04251 Leipzig, Germany
E-mail: info@ibai-publishing.org
http://www.ibai-publishing.org

14th International Conference on Mass Data Analysis of Images and Signals in Artificial Intelligence and Pattern Recognition, MDA 2019

with Applications in Medicine, r/g/b Biotechnology, Food Industries and Dietetics, Biometry and Security, Agriculture, Drug Discover, and System Biology

July 14 - 15, 2019, New York, USA

Chair

Petra Perner
Institute of Computer Vision and Applied Computer Sciences, IBaI, Germany

Program Committee

Telemedicine Data Flow Models

Calin Ciufudean

„Stefan cel Mare" University, 9 Universitatii str.,
720225, Suceava, Romania
calin@eed.usv.ro

Abstract. For medical specialists prevention is always preferred to treatment, especially when access to patient is difficult and in emergency cases it is mandatory to have a specialized intervention by distance, i.e. one may need a telemedicine facility. In order to prevent and to intervene using telemedicine staff we need good logistics, both for data transmission and good prediction tools. Our paper deals with telemedicine`s availability diagnosis by using discrete event models. Physicians' expertise in medical examination and laboratory analysis are here modeled using Markov chains and their dynamics on medical diagnosis and treatment is estimated. As Petri nets (PN) and Markov chains are well established formalisms for modeling and representing knowledge dynamics we use them for improving the state of the art in models by means of risk estimation availability, interoperability, and prevention throughout tele-medical techniques. A theoretical example will emphasize our approach.

Keywords: Telemedicine, Markov chains, pulse width modulation, rare events, discrete event systems.

1 Introduction

Environmental issues are already visible everywhere and have dangerous consequences for human health, and this is a clear sign that environmental issues entropy has reached the maximum value. We estimate the amount of logistics involved in telemedicine process in order to find the optimum scheduling of medical procedure in order to deliver an efficient diagnosis, treatment for patients, and for medical staff to save time, energy and to respect the environment. In the last century, the entropy had been widely used both in scientific research and for solving technical issues. In our approach, a tele-medical system is treated as a discrete event system and we consider that the evolution of this system can be modeled and analyzed using stochastic process formalisms. In particular, we deal with Markov chain models, and we assume the entropy of the model's states will determine the system's trajectory and dynamic. Different analytic or graphic models might be used in order to accomplish the above mentioned analysis [1-4]. Choosing a particular modeling formalism for analysis of such complex systems is due to the following reasons: first comes the nature of phenomena which have to be modeled, second is the desired representation kind, and in the third place is the user`s abilities and availabilities. An example for applying knowledge representation based on expert`s reasoning with uncertainty is applied on Markov chain dynamics [2, 3]. Our work synthesis, in a certain measure, the prestigious applied researches debated in literature [5-10]. We do not try to find a specific application of our approach, but to deal with general models. Therefore we propose a new Markov chain model, based on a Petri net support for modeling causal relations between the factors that determines the necessity and opportunity of using telemedicine diagnostic procedures, for estimating the probability of both their occurrence and impact on human health. We choose the Markov model considering that prediction and estimation of medical performance impact is a hard task involving social, economic aspects, as well as medical strategies in order to build a model suitable to manage with these interacting parameters. Typical, or new medical applications can benefit of this general model we introduce here.

The rest of the paper is organized as follows: we give a short description of the Petri nets models for controllers of tele-medical systems problem in section 2, and a design of the proposed Markov chain model for performing efficient diagnoses in telemedicine is presented in section 3. A theoretical illustrative example of our discrete event model for telemedicine systems is given in section 4. Section 5 gives some concluding remarks.

2 Petri Net Model for Telemedicine Diagnosis

We will assume that the reader is familiar with Petri nets (PN) theory and their application to applied systems, such as telemedicine, or we refer the reader to [1-4]. In a high volume transfer line (i.e., in a telemedicine environment) the global PN model should be divided into modules in order to make possible analysis. Each module is analyzed and optimized, and then the Petri nets models for modules are synchronized in order to be integrated in one global Petri net [5]. The modules preserve the structure of the entire net model, and synchronization is ensured by transitions which have physical meaning. The global Petri net model of the tele-medical system has properties of a modular logic controller. These modules of a PN controller are bided using reduction rules to reduce the complexity of the global Petri net; for example, non-synchronized transitions are rejected. Figure 1 displays the structure of a PN controller with three modules and three synchronized transitions. As a general rule, the PN model should be safe, and reversible, and therefore a special attention should be paid to its topology and initial marking [3, 4]. Liveness and reversibility properties of PN can be easily analyzed using the state equation of the marked Petri nets, as we exemplify for the net depicted in Fig.1.

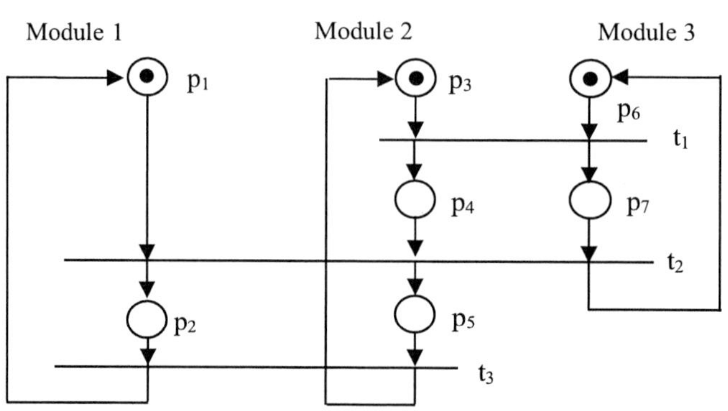

Fig.1. An example of modular PN controller.

We denote with C the incidence matrix of the global Petri net, with C_i , i = {1, 2, 3} the incidence matrix of the modules, and by M_0 the initial marking of the global Petri net, and we have:

$$C = \begin{matrix} p_1 \\ p_2 \\ p_3 \\ p_4 \\ p_5 \\ P_6 \\ P_7 \end{matrix} \begin{bmatrix} & \begin{matrix} t_1 & t_2 & t_3 \end{matrix} \\ C_1 & \begin{bmatrix} 0 & -1 & 1 \\ 0 & 1 & -1 \end{bmatrix} \\ C_2 & \begin{bmatrix} -1 & 0 & 1 \\ 1 & -1 & 0 \\ 0 & 1 & -1 \end{bmatrix} \\ C_3 & \begin{bmatrix} -1 & 1 & 0 \\ 1 & -1 & 0 \end{bmatrix} \end{bmatrix}, M_0 = \begin{bmatrix} 1 \\ 0 \\ 1 \\ 0 \\ 0 \\ 1 \\ 0 \end{bmatrix} \quad (1)$$

The initial places of the PN safe and reversible controller are represented by n-dimensional vector for the component modules:

$$M_0 \in \{0,1\}^n \quad (2)$$

Where "1" marks the initial locations of the modules, and M_0 denotes the initial marking of the PN.

An advantage of this modular construction is its versatility, i.e. it can be easily reconfigured according to different real state tele-medical systems (i.e., admit different communication protocols) by replacing incidence matrixes of modules.
The evolution of a modular PN controller is given by the next relation (state equation):

$$M = M_0 + C \cdot ft_C \quad (3)$$

Where, ft_C is the firing vector of transition "t" of the Petri net.
Data traffic flow is very important for tele-medical systems in order to ensure an operative diagnosis and treatment for patients. The traffic volume is determined by the cycle time of a telecommunication system in normal operation. As it is well known, performance analysis of PN`s is easily done by adding time specifications to

the Petri net model, i.e. to places (modeling tasks duration) and transitions (modeling data processing duration) of the PN. Throughput analysis of timed Petri nets basically means evaluation of cycle time necessary to perform a complete firing transition cycle [2 - 6].

For timed Petri nets, a well-known method for computing the minimum cycle time C_T is given by the following relation [4, 5]:

$$C_T = \underbrace{\max}_{v \in \Gamma} \left\{ \frac{D(\gamma)}{N(\gamma)} \right\} \tag{4}$$

Where Γ is the set of circuits (also named loops) of the timed Petri net;

$D(\gamma) = \sum_{p_i \in \gamma} \tau_i$ is the total delay introduced by the places of the circuit γ;

$N(\gamma)$ is the total number of tokens (resources) in the places of the circuit γ.

As mentioned in [1-3], the trajectory of timed Petri nets depends to the number of tokens and to the number of states in the main circuit (e.g. the circuit which decides the cycle time C_T), and this analysis becomes more and more difficult as the complexity of the PN model grows. Therefore, in the next section we deal with formalism capable to handle large size discrete event models, i.e. the Markov chains.

3 Markov Chain Model for Telemedicine Diagnosis

The PN model we discussed in the previous section will help us to build a Markov chain diagnosis structure.

We consider a Petri net $PN = (P, T)$ with place set $P_P = [1,...,n]$ and transition set T, with $(i, j) \in T \Leftrightarrow (j,i) \in T$. We also consider a Markov chain built with the places and transitions of the Petri net; the state at moment "t" will be denoted $S(t) \in P_P$, for $t = 0, 1, ..., n$, where $n \in N$. Transitions of the Markov chain are associated with transitions in the Petri net depicted in figure 1. The transitions probabilities must be nonnegative, and the sum of the probabilities, including self-logs, must be equal to 1 (i.e. we deal with a stochastic process). We define the probability of self-logs (i, i) as the probability that S(t) stays at place i.

The transition probability matrix $P_{ij} \in R^{nxn}$ has the following structure:

$$P_{ij} = \text{prob}\,(S(t+1) = j / S(t) = i) \quad i, j = 1,...,n \tag{5}$$

Where the stochastic matrix P_{ij} has the following properties:

$$P_{ij} \geq 0, \quad P_{ij} = P^T \tag{6}$$

From relation (7) we deduce that matrix P also satisfies the following condition:

$$P_{ij} = 0\,,\ (i,j) \notin T \tag{7}$$

Relation (7) states that transitions are allowed only between connected states. Let $\prod(t) \in R^n$ be the set of probabilities of states in the Markov chain at moment t: $\prod_i(t)$ = prob(S(t) = i), and their distribution is given by the following relation:

$$\Pi_i(t)^T = \Pi_i(0)^T \cdot P_{ij}{}^t \tag{8}$$

For the elements of $\prod_i(t)$ we have the following relation [10, 11]:

$$\pi_0 = 1;\ \pi_i = \frac{p_{01} \cdot p_{12} \cdots p_{i-1i}}{p_{10} \cdot p_{21} \cdots p_{ii-1}},\quad i \geq 1 \tag{9}$$

Considering that $\sum_i \left[\pi_i + (p_{ii+1} \cdot \pi_i)^{-1}\right] = \infty$, the following connections are true [12, 13]:

If $p_{00} = 0$ then we have a negative random circuit recurrent if and only if $\sum_{i=0}^{\infty} p_i \cdot \pi_i = \infty$, and nonnegative recurrent if $\sum_{i=0}^{\infty} p_i \cdot \pi_i \langle \infty$.

4 Illustrative Example

Let us now apply the above discussed approaches to a telemedicine system, and therefore we assume that data flow respect a Poisson process of rate β/(data slot), and the time is divided according to these slots. The duration of a data transmission is bigger than the maximum nominal processing time d_j of processing units PU_j, j = 1,..., n+1, and only one data package can be transmitted in one slot. If more than one data slot attempts to access PU_{n+1} the transmission stops (i.e. there is a bottleneck) and data should be retransmitted in subsequent slots. Slots than have been blocked are called backlogged slots. Each slot which is backlogged attempts, independently from other backlogged slots to use the following slot with retransmission probability "p". We mention that this approach is independent of the arrival process. Denote by S(n) the number of backlogged slots at moment t_n. Then S= {S(t_n), n = 0, 1,...} in the Markov chain built as we described in the previous section, and we consecutively have [8, 9]:

$$p_{ii-1} = i \cdot p(1-p)^{i-1} \cdot e^{-\beta},\qquad i \geq 1 \tag{10}$$

$$p_{ii} = \left[1 - i \cdot p(1-p)^{i-1}\right] \cdot e^{-\beta} + (1-p)^i \cdot \beta \cdot e^{-\beta},\quad i \geq 0 \tag{11}$$

$$p_{ii+1} = \left[1 - (1-p)^i\right] \cdot \beta \cdot e^{-\beta},\qquad i \geq 0 \tag{12}$$

$$p_{ii+k} = \frac{e^{-\beta} \cdot \beta^k}{k!}, \qquad i \geq 0, k \geq 2 \qquad (13)$$

$$p_{ij} = 0, \qquad \text{otherwise} \qquad (14)$$

As it is well known, the blockage of the data flow in a tele-medical system occurs when the number of backlogged parts increases in an uncontrolled manner, so that one may say that the trough put of the system will decrease to zero.

To explain this phenomenon encountered as well in communication systems [13, 14] we introduce d(i) as the upward drift of the number of backlogged parts in state i:

$$d(i) = E\{S(n+1) - S(n) \mid S(n) = i\}, i \geq 0 \qquad (15)$$

From (11) and (12) we have:

$$d(i) = \beta - c(i), i \geq 0 \qquad (16)$$

Where
$$c(i) = e^{-\beta} \cdot i \cdot p(1-p)^{i-1} + \beta \cdot e^{-\beta}(1-p)^i \qquad (17)$$

It is shown in [15, 16] that d(i) attains its minimum at $i_{min} = \frac{(1-\beta)(1-p)}{p}$, where c(i) increases on $[0, i_{min}]$ and decreases on $[i_{min} + 1, +\infty]$. It is also shown in [17, 18] that if $\beta < c(i_{min})$ then we define states as stable, respectively unstable: i_s respectively i_n, where $(0 \leq i_s \leq i_{min} < i_n)$:

$$c(i_s) \leq \beta < c(i_s + 1) \qquad (18)$$

$$c(i_u) \leq \beta < c(i_u + 1) \qquad (19)$$

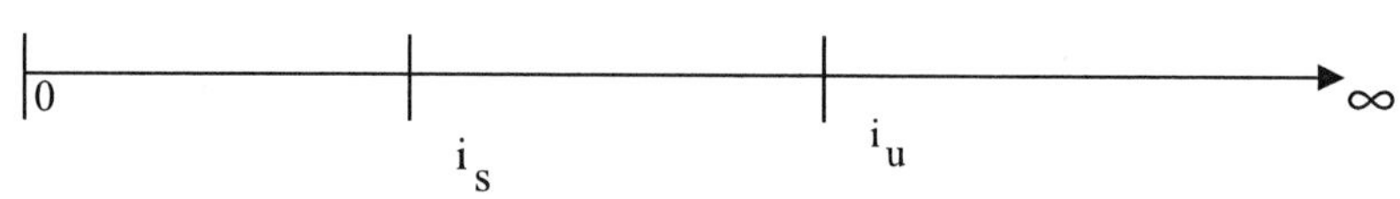

Fig. 2. Drift evolution of backlogged data slots.

In order to increase the throughput of the tele-medical system one may need to have a greater probability that the Markov chain stays as closes as possible to 0, e.g. the drift is negligible. Based on [16-18] it is easy to prove that when Markov chain is beyond state i_u, then b_i, the probability the Markov chain will reach state 0 after that line point is decreasing in i_u less than 0,5 but these probabilities cannot be neglected. In [19-21] is defined the stationary distribution as the limit, considering the number of data slots increases very much, of the conditional distribution of the Markov chain S(n) given that the throughput of the tele-medical process has not been beyond i_u [21, 22]:

$$v_i = \lim_{n \to \infty} P_\omega\{S(n) = i / S(m) \in \{0, 1, \ldots, i_n\}, \quad 0 \le m \le n-1\}, \qquad \forall i = 0, 1, \ldots, i_n \tag{20}$$

The set **V** = (v_i) shows the distribution of the backlog data slots given that the system has been operating satisfactory for a long time.

From [23, 24] we know that the limits v_i exist, are independent of the initial distribution, ω and satisfy the following relation:

$$v_i = \frac{a_i}{\sum_{k=0}^{i_u} a_k}, \qquad \forall i = 0, 1, \ldots, i_u \tag{21}$$

Where a = (a_i) is the left eigenvector of the elements $((i_{u+1}) \times (i_{u+1}))$ of P_{ij}, see relations (5), and (6), associated with its largest eigenvalue S_{i_u+1} that is, a is solution to the equation:

$$S_{i_u+1} \cdot a = a_{(i_u+1)} \cdot P_{ij} \tag{22}$$

From [25-27] we have that: $a_i = Q_i(S_{i_u+1})$, i = 0,1,..., i_u:

$$v_i = \frac{Q_i(S_{i_u+1})}{\sum_{k=0}^{i_u} Q_k(S_{i_u+1})}, \qquad i = 0, 1, \ldots, i_u \tag{23}$$

Where $Q(S_{i_u+1})$ the left eigenvector of is $_{(m)}P_{ij}$ associated with the largest eigenvalue S_{i_u+1}.

As for a pure theoretical example, let us assume a_5 = 0.35 and p = 0.4, and then we have i_u = 5, and $b_6 = \lim_{m \to \infty} b_5^{(m)} = 0{,}337$. It results that quasi-stationary distribution provides a well enough limit that characterizes the stable behavior of the backlog operation.

5 Conclusion

In this paper, we presented few theoretical considerations of basic discrete event models of data flow of tele-medical systems. Our models use Petri nets and Markov chains formalisms. Performance metrics for the diagnosis models of data flow were also defined. These diagnosis models can be easily incorporated in a user friendly controller modeled with Petri nets as they are related to Markov chains. The theoretical illustrative example demonstrated the effectiveness of our approach. Due to intuitive interface between man and machine we consider these models suitable for development in tele-medical applications in order to have a versatile tool for telemedicine. We also believe that this paper delivers several paths for future works for rare events` formalisms diagnosis of security communication system based on finite-state, discrete event parameter, recurrent Markov chain.

References

1. Aeschliman, R.: Modelling and Analysis of Tele-mental Health Systems with Petri Nets, M.S. Thesis, Kansas State University, (2015).
2. Mtibaa, S., Tagina, M.: An Automated Petri-Net Based Approach for Change Management in Distributed Telemedicine Environment, www.researchgate.net/publication/232608258, (2012).
3. Murata, T., Petri nets: Properties, analysis and applications, in Proc. IEEE, pp. 541-580, (1989).
4. Papastergiou, G, s.a.: Deep-Space Transport Protocol: A Novel Transport Scheme for Space DTNs, (2009).
5. Perner, P.: Prototype-Based Classification, Applied Intelligence 28(3): 238-246 (2008).
6. Perner, P.: Case-base maintenance by conceptual clustering of graphs, Engineering Applications of Artificial Intelligence, vol. 19, No. 4, pp. 381-295 (2006).
7. Kirkizlar, E., Serban, N., s.a.: Evaluation of Telemedicine for Screening of Diabetic Retinopathy in the Veterans Health Administration, Ophthalmology Volume 120, Issue 12, pp. 2604-2610, (2013).
8. Nguyen, H.V., Tan, G.S., s.a.: Cost-effectiveness of a National Telemedicine Diabetic Retinopathy Screening Program in Singapore, Ophthalmology, pp. 123(12): 2571-2580, (2016).
9. Kumar, P., Sharma, S.K., Prasad, S.: CAD for the Detection of Fetal Electrocardiogram through Neuro-Fuzzy Logic and Wavelets Systems for Telemetry, Second International Conference on Computational Intelligence & Communication Technology (CICT), pp. 121-127, (2016).
10. Llamedo, M., Martin-Yebra, A., s.a.: Non-invasive FECG estimation based on linear transformations, *Computing in Cardiology Conference*, pp. 285-288, 2013.
11. John A. Brennan, J.A., Krohmer, J.R.: Principles of EMS Systems, Third Edition, (2006).
12. Cruz-Cunha, M. M., s.a. : Encyclopedia of E-Health and Telemedicine, IGI Global, (2016).
13. Ermonand, S., Gomes, C.P, s.a.: Designing Fast Absorbing Markov Chains, Associationfor the Advancement of Artificial Intelligence, https://cs.stanford.edu/~ermon/papers/aaai14-mcmc.pdf, (2014).
14. Ciufudean, C.: "Risk and Reliability Analysis of Flexible Construction Robotized Systems", Robotics and Automation in Construction, INTECH, 2008.
15. Ciufudean, C.: Discrete Event Frameworks of Environmental Sustainable Development, Lap Lambert, Germany, (2013).
16. Ciufudean, C.: Discrete Event Models in Telemedicine, Advances in Mass Data Analysis of Images and Signals, 12th International Conference, MDA 2017, New York, USA, pp.44-54, (2017).
17. Ciufudean, C.: Discrete Event Systems Applied in Medicine: Formalisms, Methods, Applications, Lap Lambert, Germany, (2011).

18. Chen, Yaming: Linear least square method for the computation of the mean first passage times of ergodic Markov chains, www.researchgate.net/publication/327010381/Linear least square method for the computation of the mean first passage times of ergodic Markov chains, (2018).
19. Brehends, E.: Introduction to Markov Chains with Special Emphasis on Rapid Mixing, Advances Lectures on Mathematics, Vieweg, Germany, (2000).
20. Karlin, S., McGregor, J.L.: Random Walks, Illinois J. Math., no.3, pp.66-81, (1959).
21. A.S. Lewis, "Nonsmooth Analysis of Eigenvalues", Mathematical Programming, no.84, pp.1-24, (1999).
24. Athreya, K.B., Ney, Branching Processes, Springer Verlag, Berlin, (1972).
25. 26. Benson, S., Ye, Y., and Zang, X.: Solving large-scale sparse semi-definite programs for combinatorial optimization", SIAM Journal Optimization, no.10, pp.443-461, (2000).
27. Koller, G., Raidl, G.R.: An evolutionary algorithm for the maximum weight trace formulation of the multiple sequence alignment problem, Lecture notes in computer science, vol. 2463, pp.40-52, Berlin, Springer Verlag, (2002).

Image analysis technique and procedure for its application in 2D

Javier Bilbao[1], Imanol Bilbao[1], Andoni Olozaga[1] and Cristina Feniser[2]

[1] University of the Basque Country, Engineering School, Pl. Ing. Torres Quevedo, 1, 48013 Bilbao, Spain
[2] Technical University of Cluj Napoca, Faculty of Machine Building, 28 Memorandumului street, 400114 Cluj-Napoca, Romania
javier.bilbao@ehu.eus

Abstract. The amount of data stored at the present is growing exponentially. Big Data is one of the research areas that now has the most future for the treatment of such a huge amount of data. At the same time, Data Mining is a fundamental part in the treatment of these data and in the search for special characteristics that may be interesting for the researcher. In the field of Medicine, the use of images for the diagnosis and analysis of certain diseases or clinical processes is nowadays a common practice. The semi-automation or automation of image processing is an area that is open to the use of different models and techniques, depending on the type of image, quality, disease, etc. The Principal Component Analysis is one of these techniques.

Keywords: big data, data mining, medical images, singular value decomposition, principal component analysis.

1 Introduction

The use of the so-called big data in different areas of knowledge and in different industrial sectors is, nowadays, a line of research in vogue. The mass data analysis can be applied in the treatment of images, among other fields. And within the field of treatment of images or the use of digitalization, Medicine is one of the sectors in which its use seems more evident. Already in 1854, when there was an epidemic of cholera in London, John Snow, considered the father of epidemiology, was meticulously detailing the places of affected homes. After a long time, since it was the nineteenth century, he deduced that the cholera that had spread through London was caused by the consumption of water contaminated with faecal matter from a water pump in Broad Street, and recommended that it should be closed down [1]. That work, with the digital positioning that we have, could have been done in a few hours in our time.

The computerization of all kind of processes has caused companies and organizations of all types to accumulate a huge amount of data. Therefore, new technologies are required to manage and extract the value of complex data that are generated in large volumes at high speeds. These data can be provided for several things, interconnected or not. It is the Internet of Things, which generates a huge flow of data, including images. Although this phenomenon is affecting all sectors, the health sector is one of the areas in which the incidence of this phenomenon is being especially relevant, due among other elements to the implementation of the electronic devices, to the interconnection of departments, to the recent explosion in the generation of genomic data (in part, thanks to its cheapening), and to the fact that the majority of data generated in the sector is unstructured.

With the enormous amounts of data that exist in digital format at the present, and with those we generate constantly with everyday acts (such as the use of electronic devices connected to the Internet, interaction in social networks, collaborative work in digital systems, the use of platforms for learning, among many others), added to the increase of the capacities of automated processing to costs increasingly low, it was logical that, in the evolution of the uses and applications of the Technologies of Information and Communication in the XXI century, different methods and tools were developed for taking advantage of these data, seeking their understanding and improving decision making.

But it is not only the great volume of information and the enormous variety of data that can be processed, generated both by human intervention and by the communication between the computers or by themselves. The importance lies mainly in the potential impact of the findings that may emerge from their analysis. So, we have to take into account not only the quantity and the quality of data, but also the way to manage and analyze them.

2 Big data

At this point, the term "big data" is the key for beginning to understand the image processing. Big data can be defined as the actions that can be performed on large-scale data sets, whose variety and volume exceed the capacity of traditional software for its capture, management and processing in reasonable times according to the high speed at which they produce, as much by human intervention as by the interaction between digital devices, in order to analyze them, understand them, generate hypotheses, make decisions, extract new ideas and knowledge, or create new forms of value.

Big data focuses on finding associations, patterns and trends among the data, unlike other techniques that try to find cause-effect relationships or projections based on probability. Therefore, it understands and requires techniques, algorithms and analytical approaches in conjunction with new proposals to design the architecture of the information that is processed.

According to De Mauro et al. (2016), Big Data is the Information asset characterized by such a High Volume, Velocity and Variety to require specific Technology and Analytical Methods for its transformation into Value. They suggested this definition as congruent with the most prominent features of the Big Data. But there are really a lot, where the most used can be the following: Big data is high-volume, high-velocity and/or high-variety information assets that demand cost-effective, innovative forms of information processing that enable enhanced insight, decision making, and process automation (Beyer and Laney, 2012); Big Data presents a fundamentally different approach that begins with data collection, then analysis, and then drawing conclusions from the patterns that appear. Establishing causality is, of course, desirable, but it requires expert knowledge, theory, and the testing of a hypothesis to prove results (Mayer-Schönberger and Cukier, 2013); Big Data is data that exceeds the processing capacity of conventional database systems. The data is too big, moves too fast, or does not fit the strictures of your database architectures. To gain value from this data, you must choose an alternative way to process it (Dumbill, 2013); it is the development of compact and scalable data structures and algorithms that support rapid data filtering, aggregation, and display rendering (Shneiderman, 2008); big data is a combination of four characteristics, namely: volume, variety, velocity and veracity, what creates an opportunity for organizations to gain competitive advantage in today's digitized marketplace (Miele and Shockley, 2013); big data is defined by four characteristics: volume, velocity, variety and value (Dijcks, 2013), Big data is a term describing the storage and analysis of large and or complex data sets using a series of techniques including, but not limited to: NoSQL, MapReduce and machine learning (Ward and Barker, 2013). Suthaharan (2014) re-defined big data by introducing a 3D space, C3, which is defined based on three new parameters, cardinality, continuity, and complexity.

Di Bella et al. (2018) focused on the social aspects of Big Data, saying that it can be substantially considered to be process-produced data gathered tracking peoples' activities in different real or virtual environments, and they can be distinguished from "large dataset" by the fact that the former have a high level of complexity and multidimensionality whereas the latter are merely datasets with many records.

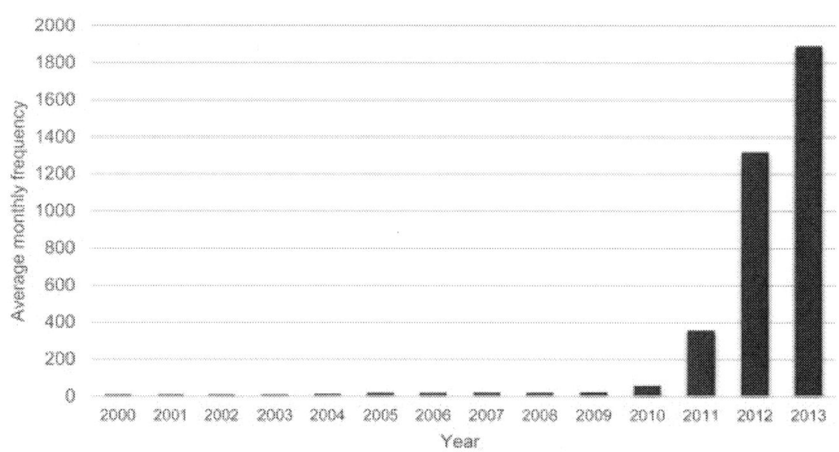

Fig. 1. Frequency distribution of documents containing the term "big data" in ProQuest Research Library (Gandomi and Haider, 2015).

From a business solution perspective, Sun et al. (2018) define big data as a new technology that is primarily characterized by its advanced business intelligence and analytics (BI&A) function. 49. Boubeta-Puig et al. (2014) considered it as an approach or type of business analytics that assists organizations to analyze large amounts of data in a timely fashion. From a technical and business point of view, Jiao et al. (2013) and Liu (2013), considered that big data could be generalized as the ever-increasing information flow from different sources that is very "big" to process.

In the field of medicine, big data technology can be applied in clinical decision making, disease monitoring, public health and research. The huge volume of existing healthcare data includes personal medical records, medical images, clinical trial data, registrations, genetic data, genomic sequences of population data, etc. More recently, this exponential growth is fueling by the generation of 3D images, as well as readings from biometric sensors or wearables. Fortunately, advances in data management, particularly virtualization and Cloud Computing, are facilitating the development of platforms for more efficient capture, storage and handling of these large volumes of data (Feldman, Martin, & Skotnes, 2012).

3 Data mining

The growth of the "digital universe" has popularized Data Mining and Big Data concepts. Both are related to the use of large data sets to handle the collection or presentation of data that serves businesses or other recipients. However, the two terms are used for two different elements of this type of operation. Big Data is a term for a large data set. Data Mining, on the other hand, refers to the activity of going through large data sets to find pertinent or timely information; this type of activity is really a good example of the old axiom "looking for a needle in a haystack".

Big Data refers to the storage of large amounts of data and the procedures used to find repetitive patterns within that data. It is the technology capable of capturing, managing and processing this data in a considerable time and efficiently.

On the other hand, Data Mining is the process of identifying all the information that is relevant and is extracted from large amounts of data. The objective of this extraction is to discover patterns and trends structuring the information that has been obtained in a way that is understandable for its use.

As in any process, data mining also has to be carried out in different phases such as:

- The understanding of what is sought and the problem to be solved.
- The determination, capture and cleaning of the data needed.
- The creation of mathematical models.
- The validation and communication of the results.
- And the integration of those same results.

Data Mining brings together the advantages of several areas such as Artificial Intelligence, Statistics, Databases, Graphic Computing and Massive Processing, especially using databases as raw material.

Data Mining technology is presented as a support technology, since our capacity to store data has grown in recent years at unstoppable speeds, but the ability to process and use those data has not gone at the same time. Data Mining is a good resource to extract value from Big Data.

Several authors have tried to define Data mining in different ways during the last years. It is used as a tool to realize automatic collection, automatic transmission, integrated query and analysis via integrating and appraising information from customers; and it is also called as Knowledge Discovery in Database (KDD) (Turban et al., 2007). Moreover, Bose and Mahapatra (2001) defined data mining as a process of identifying interesting patterns in databases, which can be used in decision making. Data mining can also be defined as a process that uses statistical, mathematical, artificial intelligence, and machine learning techniques to obtain and identify valuable information and subsequently gain knowledge from a large database (Turban et al., 2007).

More recently, other authors such as Rosli et al. (2018) and Shankar (2017) approximate data mining as a process that uses statistics, mathematical, artificial, and machine learning techniques to extract and identify useful information and related knowledge from various database.

Witten et al. (2017) provide the best definition when they say that Data Mining is the process of discovering patterns in data, where this process must be automatic or (more usually) semiautomatic, and where the patterns discovered must be meaningful in that they lead to some advantage, e.g. an economic advantage.

The use of Computer Science and ICT in the field of Health is entering a new era where technology is beginning to handle large volumes of data, leading to unlimited potential for information growth. Data mining and large data analysis (related to Big Data) are helping to make decisions regarding diagnosis, treatment, etc. And everything finally focused on better patient care. For example, the use of data mining in health in the United States can save the health industry up to 450 billion dollars each year (Kayyali, Knott, & Van Kuiken, 2013). This is due mainly to the increasing volumes of generated data and the technologies to analyze them.

The explosive growth of data, already in the '80s, generated the appearance of a new field of investigation that was called KDD (Knowledge Discovery in Databases). Under this acronym, the knowledge discovery process is hidden in large volumes of data (Fayyad, Piatetsky-Shapiro, & Smyth, 1996). Nowadays, general public knows it as Data Mining.

4 Data Mining applied to medical images

One of the great challenges of Data Mining is the development of new algorithms and methodologies for the processing, analysis and interpretation of the enormous volume of data that images represent, particularly medical images, in order to help health professionals to exploit all the information contained in them. Since the extraction of information from these images depends on many factors, such as modalities, registration conditions, devices, etc., it is not possible to have general procedures to locate or identify objects such as anatomical structures or injuries, but it is necessary to develop particular methods for each type of image or pathology.

The medical image is today one of the main sources of information used by doctors for diagnosis and therapy. They can be considered a representation of the human body, which, in a practically innocuous way, allow the characterization of different diseases, facilitating their diagnosis and treatment. As a result, all technologies related to medical imaging have evolved significantly in recent decades.

The software for image processing has expanded including sophisticated algorithms for, among other objectives:

- highlight specific characteristics of the original image,
- manipulate the presentation of the image,
- correct the distortions caused by the acquisition equipment, and
- perform other mathematical analyzes in order to extract information of diagnostic utility.

This variety of operations is feasible due to having the image data stored in numerical form, which allows their mathematical processing.

One of the most commonly used method of image processing for recognition of special characteristics in one image is the Principal Component Analysis (Turk and Pentland, 1991a; Kirby and Sirovich, 1990; Turk and Pentland, 1991b; MIT, 2002; Ma and Aybat, 2018; Peng et al., 2019), which is based on the mathematical properties of the digitized image, which captures invariant characteristics of the images. It is interesting to study and analyze this technique for the following reasons:

- Apparent simplicity of implementation against good results in large databases.
- It is a technique resistant to variations.
- It is carried out under a purely automatic process.

The Singular Value Decomposition (SVD) is a very effective technique of matrix decomposition, and it can be used to solve different types of mathematical problems (sets of linear equations, problems of least squares, etc.). We used in the image processing.

The singular value decomposition is defined as:

Given a non-square matrix A of dimensions n x m and of rank r, there are orthogonal and unitary matrices U_{nxn} and V_{mxm}, and a positive diagonal matrix S_{rxr}, such that

$$A = U\Sigma V^T$$

where

$$\Sigma = \begin{bmatrix} S & 0 \\ 0 & 0 \end{bmatrix} \; ; \; S = \begin{bmatrix} \sigma_1 & 0 & \cdots & 0 \\ 0 & \sigma_2 & \cdots & 0 \\ \vdots & \vdots & \ddots & \vdots \\ 0 & 0 & \cdots & \sigma_r \end{bmatrix}$$

In Matrix Algebra, the matrices U and V represent a simple change of coordinates and have a close relationship with the eigenvectors and eigenvalues of AA^T and A^TA, respectively.

Matrix A has min (m, n) singular values, denoted by σ_i, r of which singular values are unique and distinct from zero. Normally, the algorithms that calculate the singular value decomposition of a matrix offer these in decreasing order according to their magnitude:

$$\bar{\sigma} \equiv \sigma_1 \geq \sigma_2 \geq \cdots \geq \sigma_r \equiv \underline{\sigma}$$

The structure of the matrix Σ allows the matrix A to be rewritten as:

$$A = \sigma_1 \cdot u_1 \cdot v_1^T + \sigma_2 \cdot u_2 \cdot v_2^T + \ldots + \sigma_r \cdot u_r \cdot v_r^T$$

where u_i and v_i are the i-th columns of the matrices U and V, respectively. The above equation illustrates the close relationship between a singular value σ_i and the corresponding columns of matrices U and V.

With the singular values, we can also calculate the condition number of a matrix ($k(A)$), which is defined as the ratio between the largest and the smallest singular values:

$$k(A) = \frac{\bar{\sigma}}{\underline{\sigma}}$$

and it is an indicator of the bad (or good) conditioning of a matrix.

The SVD decomposition has excellent numerical properties allowing the reliable determination of the rank of a matrix and the calculation of pseudo-inverses.

The pseudo-inverse A^+ of a matrix A, is defined as:

$$A^+ = V\Sigma^+U^T$$

where

$$\Sigma^+ = \begin{bmatrix} S^{-1} & 0 \\ 0 & 0 \end{bmatrix} \; ; \; S^{-1} = \begin{bmatrix} \frac{1}{\sigma_1} & 0 & \cdots & 0 \\ 0 & \frac{1}{\sigma_2} & \cdots & 0 \\ \vdots & \vdots & \ddots & \vdots \\ 0 & 0 & \cdots & \frac{1}{\sigma_r} \end{bmatrix}$$

The singular value decomposition is a mathematical tool widely used in multivariable control theory. For example:

- The SVD decomposition is applied to the frequency response of a system to obtain information about its gain and the main directions of the system (Skogestad and Postlethwaite, 1996).

- In the theory of robust control, SVD decomposition is used to obtain margins of robustness of multivariable systems against modeling errors (Maciejowski, 1989).
- It is useful in techniques for reducing the order of linear models (Balanced realization). Commonly, linear models in space of states obtained from non-linear models include a large number of states, most of which do not influence too much the transfer functions of interest. Using the SVD decomposition, we can identify which states can be eliminated without introducing large errors in the resulting model (Zhou and Doyle, 1998).
- It can be used directly as a decoupler (Steady State Decoupler) in order to control the process as a set of simple and independent loops, to determine the best position of the sensors, or to decide which the best manipulated variables are to control the process (Deshpande, 1989).

5 Principal Component Analysis

In Engineering, a large number of problems that handle huge amounts of numerical data are represented in the form of matrices and their solution is to perform calculations with these matrices. Because these data contain a large amount of redundant information, problems of bad numerical conditioning or collinearity may appear. Then, it would be better to be able to separate this redundancy from the data, either selecting the key variables or applying methods to reduce the dimensionality of these. Other times, the problem is not a "redundancy", but simply a non-interesting data, like in an image of some lungs, where what is wanted to be discerned is if there is some type of cancer in some point of those organs.

The Principal Component Analysis (PCA) has been widely used for this purpose and is generally one of the most widespread method for extracting common information from large amounts of data (Abdi and Williams, 2009).

The analysis of main components makes use of the SVD decomposition, and geometrically it can be described as a projection of the original data on a different hyperplane, that of the main components. This transformation reveals the relationships between the different variables that make up the matrix. Each main component explains a percentage of the variation of the data, the first being the most important, then the second, and so on.

If these variables are correlated or redundant, there will be rows or columns dependent on each other in such a way that the number of singular values other than zero will be equal to the rank of the matrix. Thus, the original data can be represented by a small number r of main components and, therefore, it is possible to use only that number of components to represent the information. In this way, the PCA technique reduces the dimensionality of the original data matrix and, with only a small set of components, most of the variability of the data can be expressed (Artoni et al., 2018)).

The analysis of main components has found application in many disciplines (statistical analysis, data compression, etc.). For example, in process control, the PCA technique has been widely used for the detection and isolation of perturbations (Ku et al.,

1995); monitoring based on statistical control (SPC) (Nomikos and MacGregor, 1995), (MacGregor and Kourti, 1995), (Thomas et al., 1996); detection and diagnosis of failures (Lewin, 1995), (Haiqing et al., 2000); and modeling and control of multivariable processes (Shah et al., 1998), since it allows easily the extension of the principles of statistical monitoring of processes to multivariable processes. And in the image processing field applied to Medicine (Upadhyaya et al., 2019).

6 Application to 2D images

We have used medical images in 2D, which have been associated with the pixels that represent different levels of gray color, from white to black (Fig. 2). These values can represent bones, cartilages, muscles or other type of tissue, including zones with some type of disease.

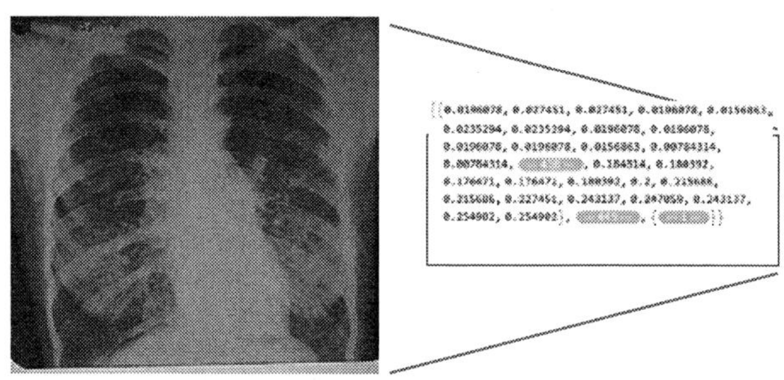

Fig. 2. Digitalization of an image: part of the data matrix.

The images in Figure 3 show the same lungs but with a different resolution, although the physical size of the image is the same. That is, using the SVD technique, we can go from having a 451x466 pixel image that weighs 37000 bytes to having an image of 451x466 pixels that weighs only 18000 bytes, without it being perceptible that the resolution of the image has worsened.

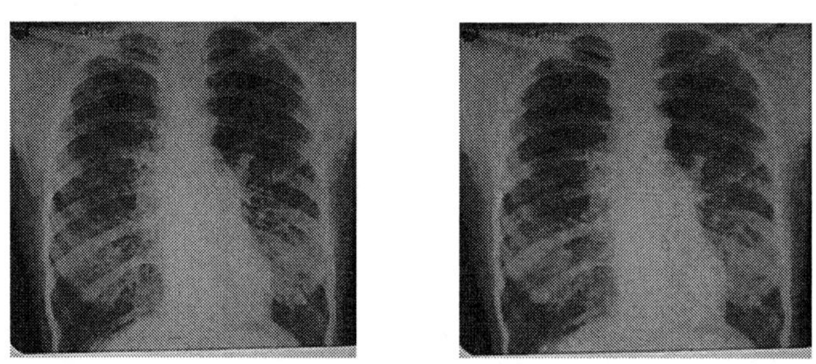

Fig. 3. Original image (left), with 37.953 bytes; and image after selecting the first 40 eigenvalues (right), with a weight of 18.691 bytes

The number of eigenvalues calculated from the image was 466 (obviously, due to its dimension), but when we represent them graphically, we see that the size of those eigenvalues decrease exponentially very fast. In Fig. 4, we show the first 100 eigenvalues, trunked at the beginning for a better representation (notice that the first value is 38763 and the 9th value is 53,0513). This graphic can give us an idea of the number of eigenvalues that we have to take into account.

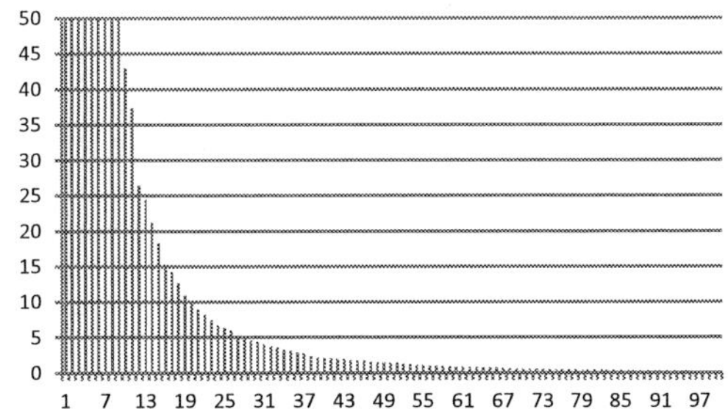

Fig. 4. First 100 eigenvalues

In the process, we can select a specific area of the image where we suspect that something can be (Fig. 5). This focus on a smaller area normally takes time to the specialist but save calculation time if the image is very big.

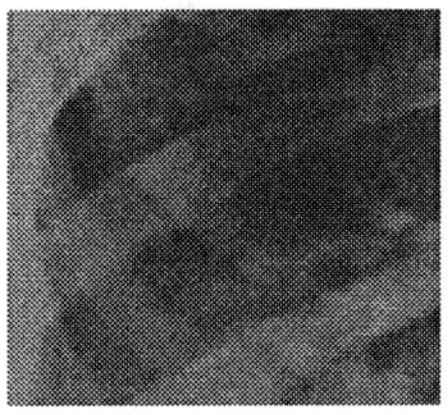

Fig. 5. Detail of the image of the lungs.

Principal Component Analysis can be adapted for the multivariate data analysis in many ways. PCA provides an approximation of a data matrix, such as an image, considering it as an A data matrix with dimensions m x n -where the m rows are the cases and the n columns, the variables- which is expressed as the product of two matrices F and G that best represent their essential patterns. In this way, we can write:

$$A = F \cdot G^T + E$$

where F are the scores (m x q) and G the loadings (n x q), coefficients of the variables, and where q, the number of Principal Components ($q <= \min(m, n)$). In the PCA model, the importance of a variable is indicated by its variance, while the matrix E contains the part of the data not explained by the model (normally called residuals).

The variance explained by each component is the associated algebraic concept of eigenvalue. The assumption in PCA is that the vectors of scores and loadings that correspond to the largest eigenvalues contain the most useful information that is related to the problem, so their eigenvectors are usually represented in descending order. It is the interpretation of eigenvalues, which give "the importance" of the variable, as we have shown in the previous paragraphs and figures of the lungs.

We implemented the algorithm of the Singular Value Decomposition to realize the PCA model. We can find the eigenvalues according to the number of the main component. Then, we have to choose - or let it be calculated automatically - the number of main components, according to the proportion of variance described by each eigenvalue or based on several criteria that exist to make this selection.

7 Conclusions

The volume of data at a general level, and in the field of medicine in particular, has skyrocketed in recent years. The number of images saved is increasing, and the quality of them is better, so any method of processing these images is more expensive from the point of view of calculation time and calculation method.

Within the various methods that exist for the treatment of images, in this article we have focused on the Singular Value Decomposition together with a later Principal Component Analysis (PCA).

PCA can become another way of handling badly conditioned problems, since it produces suboptimal solutions but with small changes in control actions. The analysis of the main components, making use of the singular value decomposition can be presented as a valid method for the processing of images.

References

1. Abdi, H., & Williams, L. J. (2009). Principal Component Analysis. In: Li S.Z., Jain A. (eds) Encyclopedia of Biometrics. Springer, Boston, MA.
2. Artoni, F., Delorme, A., & Makeig, S. (2018). Applying dimension reduction to EEG data by Principal Component Analysis reduces the quality of its subsequent Independent Component decomposition, NeuroImage, 175, 176-187.
3. Beyer, M.A. and Laney, D. (2012), The Importance of "Big Data": A Definition, Gartner, Stamford, CT.
4. Bose, I., & Mahapatra, R. K. (2001). Business data mining-a machine learning perspective. Information Management, 39(3), 211-225.
5. Boubeta-Puig J, Ortiz G, Medina-Bulo I. A model-driven approach for facilitating user-friendly design of complex event patterns. Expert Systems with Applications 41(2):445–456 (2014).
6. De Mauro, A., Greco, M., Grimaldi, M.: A formal definition of Big Data basedon its essential features. Library Review 65(3), 122–135, (2016). https://doi.org/10.1108/LR-06-2015-0061
7. Deshpande, P.B. (1989). Multivariable Process Control. Instrument Society of America.

8. Di Bella, E., Leporatti, L. and Maggino, F.: Big Data and Social Indicators: Actual Trends and New Perspectives. Social Indicators Research 135(3), 869–878 (2018) https://doi.org/10.1007/s11205-016-1495-y
9. Dijcks, J.P. (2013), "Oracle: Big data for the enterprise", Oracle White Paper, Oracle Corporation, Redwood Shores, CA.
10. Dumbill, E. (2013), "Making sense of big data", Big Data, Vol. 1 No. 1, pp. 1-2.
11. Fayyad, U. M., Piatetsky-Shapiro, G., & Smyth, P. (1996). From data mining to knowledge discovery: an overview. In U. M. Fayyad, G. Piatetsky-Shapiro, P. Smyth, & R. Uthurusamy (Eds.), Advances in Knowledge Discovery and Data Mining (pp. 1–34). Menlo Park, CA, USA: American Association for Artificial Intelligence. Retrieved from http://dl.acm.org/citation.cfm?id=257938.257942
12. Feldman, B., Martin, E.M. & Skotnes, T.: Big Data in Healthcare. Hype and Hope. (2012). Available at: http://www.west-info.eu/files/big-data-in-healthcare.pdf
13. Gandomi, A., & Haider, M. (2015). Beyond the hype: Big data concepts, methods, and analytics. International Journal of Information Management, 35, 137-144.
14. Haiqing, W., S. Zhihuan and L. Ping (2000). A novel approach to multivariate statistical process monitoring and diagnosis. In: 3rd. Asian Control Conference. Shanghai. pp. 2602–2606.
15. Jiao Y, Wang YH, Zhang SH, Li Y, Yang BM, Yuan L. A cloud approach to unified lifecycle data management in architecture, engineering, construction and facilities management: integrating BIMs and SNS. Adv Eng Inf. 2013;27(2):173–188.
16. Kayyali, B., Knott, D., & Van Kuiken, S.: The big-data revolution in US health care: Accelerating value and innovation. In McKinsey Company. McKinsey Company. (2013). Available at: http://www.mckinsey.com/insights/health_systems_and_services/the_big-data_revolution_in_us_health_care
17. Khoury, M.J., Ioannidis, J.P.: Big data meets public health. Science 346 (6213), 1054–1055 (2014).
18. Kirby, M. and Sirovich, L. "Application of the karhunen-loeve procedure for the characterization of human faces". IEEE Trans. Pattern Analysis and Machine Intelligence, 12(1):103–108, 1990.
19. Ku, W., R.H. Storer and C. Georgakis (1995). Disturbance detection and isolation by dynamic principal component analysis. Chemometrics and Int. Lab. Systems 30, 179–196.
20. Lewin, D.R. (1995). Predictive maintenance using PCA. Control Eng. Practice 3(3), 415–421.
21. Liu L. Computing infrastructure for big data processing. Front Comput Sci. 2013;7(2):165–170.
22. Ma, S., & Aybat, N. S. (2018). Efficient Optimization Algorithms for RobustPrincipal Component Analysis and Its Variants, Proceedings of the IEEE, 106(8), 1411-1426.
23. MacGregor, J. F. and T. Kourti (1995). Statistical process control of multivariate processes. Control Eng. Practice 3(3), 403–414.
24. Maciejowski, J.M. (1989). Multivariable Feedback Design. Addison-Wesley.
25. Mayer-Schönberger, V. and Cukier, K. (2013), Big Data: A Revolution That Will Transform How We Live, Work and Think, John Murray, London.
26. Miele, S. and Shockley, R.: Analytics: The real-world use of big data. How innovative enterprises in the midmarket extract value from uncertain data. IBM Institute for Business Value, Said Business School, New York, NY, (2013).
27. MIT Media Laboratory Vision and Modeling Group (VISMOD) Face Recognition Demo (2002). Page http://vismod.media.mit.edu/vismod/demos/facerec/index.html

28. Nomikos, P. and J.F. MacGregor (1995). Multivariate SPC charts for monitoring batch processes. Technometrics 37(1), 41–59.
29. Peng, J., Yu, K., Wang, J., Zhang, Q., Wang, L., & Fan, P. (2019). Mining painted cultural relic patterns based on principal component images selection and image fusion of hyperspectral images, Journal of Cultural Heritage, 36, 32-39.
30. Rosli, M.R.B., Salamon, H.B., and Huda, M. (2018). Distribution Management of Zakat Fund: Recommended Proposal for Asnaf Riqab in Malaysia. International Journal of Civil Engineering and Technology 9(3), pp. 56–64.
31. Shah, S., R. Miller, H. Takada, K. Morinaga and T. Satou (1998). Modelling and control of a tubular reactor: a PCA-based approach. In: IFAC DYCOPS. Corfu, Greece.
32. Shankar, K. "Prediction of Most Risk Factors in Hepatitis Disease using Apriori Algorithm. Research Journal of Pharmaceutical Biological and Chemical Sciences 8.5 (2017): 477-484.
33. Shneiderman, B. (2008), "Extreme visualization: squeezing a billion records into a million pixels", Proceedings of the 2008 ACM SIGMOD International Conference on Management of data, pp. 3-12, available at: http://doi.org/10.1145/1376616.1376618
34. Skogestad, S. and I. Postlethwaite (1996). Multivariable feedback control. Analysis and Design. John Wiley and Sons.
35. Sun, S., Cegielski, C.G., Jia, L. & Hall, D.J.: Understanding the Factors Affecting the Organizational Adoption of Big Data. Journal of Computer Information Systems, (2016) DOI: 10.1080/08874417.2016.1222891
36. Suthaharan, S. (2014). Big data classification: Problems and challenges in network intrusion prediction with machine learning. Performance Evaluation Review, 41(4), 70-73. doi: 10.1145/2627534.2627557
37. Thomas, C., T. Wada and D. E. Seborg (1996). Principal component analysis applied to process monitoring of an industrial distillation column. In: 13th. IFAC World Congress. San Francisco (USA).
38. Turban, E., Aronson, G. E., Liang, T. P., & Sharda, R. (2007). Decision support and business intelligence systems (Eighth ed.). Pearson Education.
39. Turk, M. and Pentland, A. "Eigenfaces for recognition". Journal of Cognitive Neuroscience, 13(1):71–86, 1991.
40. Turk, M. and Pentland, A. "Face recognition using eigenfaces", in Proceedings of International Conference on Pattern Recognition , pp. 586-591,1991
41. Upadhyaya, R., Pansea, P., Sonia, A., Bhatt, U. R. (2019). Principal Component Analysis as a Dimensionality Reduction and Data Preprocessing Technique, Proceedings of Recent Advances in Interdisciplinary Trends in Engineering & Applications (RAITEA) 2019.
42. Ward, J.S. and Barker, A. (2013), "Undefined by data: a survey of big data definitions", available at: arXiv:,1309.5821[cs.DB]
43. Witten, I.H., Frank, E., Hall, M.A., Pal, C.J.: Data Mining. Practical Machine Learning Tools and Techniques. 4th edn. Elsevier (2017).
44. Zhou, K. and J. C. Doyle (1998). Essentials of Robust Control. Prentice Hall.

Indirect Determination of Mycotoxin Concentration with Image Processing and Data Mining

Petra Perner

Institute of Computer Vision and Applied Computer Sciences, IBaI, Leipzig, Germany
pperner@ibai-institut.de

Abstract. We have developed a novel method for the detection of hygiene-relevant parameters from grains of cereal crops based on a novel probe handling method, intelligent image acquisition and interpretation methods as well as on data mining. We present our results that describe the data acquisition, the image analysis and interpretation method as well as the reasoning methods that can map the automatic acquired parameters of grain to the relevant hygiene parameter´s such as Mycotoxin Concentration. The results show that with the new computer science methods such as image processing and data mining it is possible to come up with new insights into the quality control of food stuff.

Keywords: Determination hygiene-relevant parameter, Mycotoxin Determination, Cereal Crop Quality Determination, Probe Handling, Image Acquisition, Image Analysis and Interpretation, Data Mining

1 Introduction

Fungal contamination of cereals is a serious economic problem throughout the world. Several fungi cause a reduction of grain quality, especially changes in color and taste [1], [2], and [3]. However, the main risk of fungal damage arises from the production of toxic compounds, known as mycotoxins. Mycotoxins can cause serious adverse health effects. Toxigenic fungi that produces mycotoxins in grains of cereals or oil seeds belong to the genera Aspergillus, Alternaria, Fusarium and Penicillium. The control of this problem is therefore of particularly interest in food safety and quality control programs.

The aim of the research is the development of a probe handling, an automatic image acquisition and image interpretation system for the fast recognition of cereal grains damaged by fungi. Thereby a data acquisition unit has been developed that allows to take the coverage from the grain and allows to place it under a microscope for the acquisition of a digital image. This image should be used in order to automatically determine the number and the kind of fungi spores contained on the grain. For suitable intelligent image analysis and interpretation methods have been developed. Based on the enumeration of fungal spore classes we can map this information to the hygiene-relevant parameters by data mining [5]. The results show that the proposed methods based on intelligent image analysis and data mining are very suitable to capture the desired information and allow to recognize formerly unknown information that can be helpful to determine the quality of food staff.

In Section 2 we describe the material used for our work. The probe handling and image acquisition is described in Section 3. Section 4 describes the intelligent image analysis and interpretation and the results for image interpretation. The mapping of image information to hygiene relevant parameter with Data Mining is given in Section 5 as well as results. Conclusions are given in Section 6.

2 Material

For our work we have been used different quality classes of wheat grains:

1. visual optical perfect grains from a charge where no fungal grains were included,
2. fungal damaged grains,
3. gall-mosquito damaged grains, and
4. visual optical perfect grains taken from a charge of fungal damaged grains.

In total we had 10 samples from each class. Thirty single grains were taken from each sample for further evaluation.

3 Probe Handling and Image Acquisition

The main problem was to make the coverage on the grains visible under the microscope and make it useable for further digital processing. Therefore, we have developed a procedure for taking the coverage from grains and bring it onto a medium that can be placed under a microscope. From there can be acquired a digital image with the help of a digital camera connected with the microscope.

The method of choice was a water-based extraction method. The grains were placed into a boil together with stones. This water-filled boil was shacked for 2 minutes, then the water was filled into a centrifuge and the sediment was put on a slide. This slide was placed under the microscope and a digital image was taken. There are other methods for extracting the coverage from the grain possible, but this should not be the main topic of this paper. The resulting digital images are shown in Figure 1a-4a.

4 Intelligent Image Analysis and Interpretation

4.1 Image Analysis

The main aim of the image analysis was to recognize possible fungi spores and process them further for further determination of the type of fungi spore. Here we used our novel case-based object recognition method [4] developed for recognizing biological objects with high variation. For the architecture of such a system see Figure 5. The case-based object recognition method uses cases that generalize the original contour of the objects and matches these cases against the contour of the objects in the image. During the match a score is calculated that describes the goodness of the fit between the object and the case. Note the result of this process is not the information what type of fungi spore is contained in the image. The resulting information tells us only if it is highly likely that the considered object is a fungi spore or not. Further evaluation is necessary to determine the kind of fungi spore. The automatically recognized objects are demonstrated in the processed images, see Figure 1b-4b. One of the main-problem of such a case-based object recognition method is to fill up the case base with a sufficiently large enough number of cases. We used our procedure described in [7] for that purpose. For the study we have 10 different cases that allow us to demonstrate the applicability of the method. The method has to be adapted to the specific image quality to show better results as well as more cases have to be learnt by our case acquisition procedure.

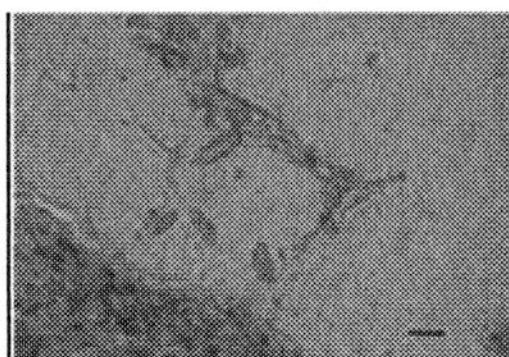	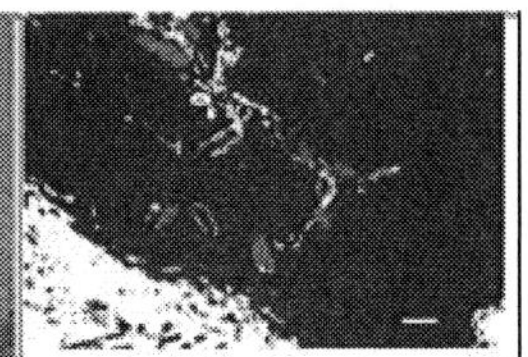
Fig. 1a Coverage of Grain with Ulocladium	**Fig. 1b** Segmented Image
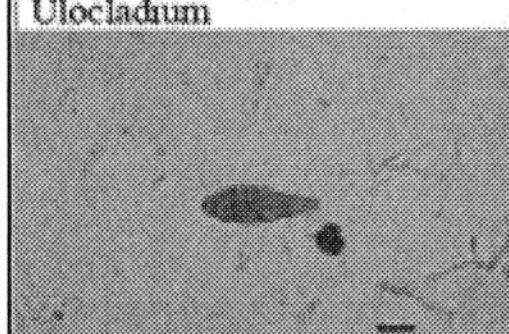	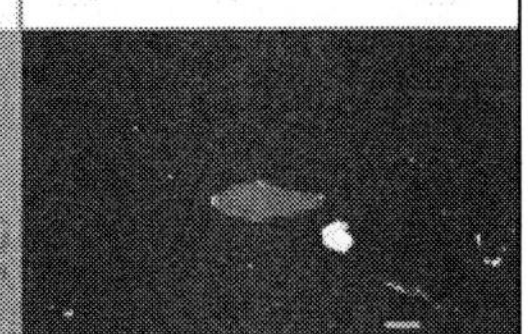
Fig. 2a Coverage of Grain with Alternaria	**Fig. 2b** Segmented Image
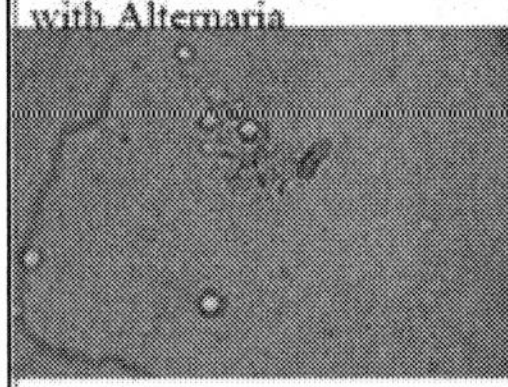	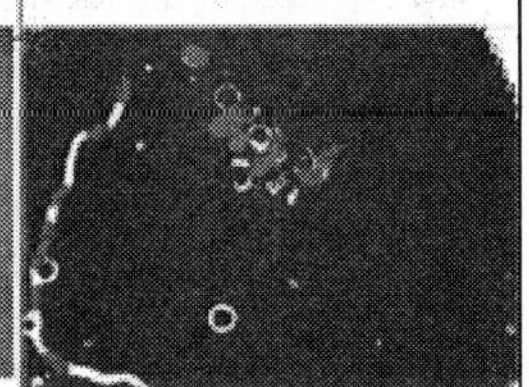
Fig. 3a Coverage of Grain	**Fig. 3b** Segmented Image
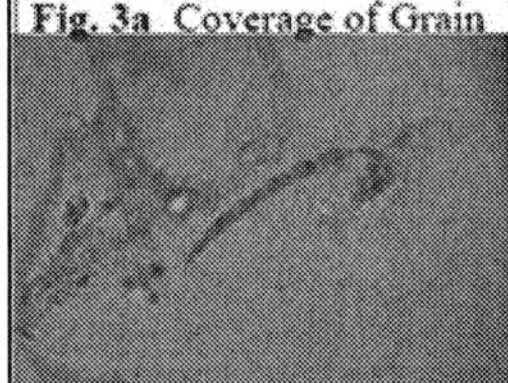	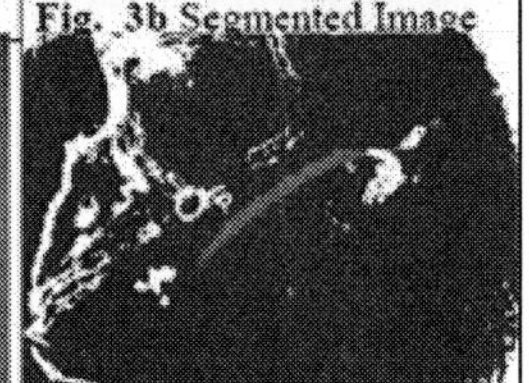
Fig. 4a Coverage of Grain with Fusarium Spore	**Fig. 4b** Segmented Image

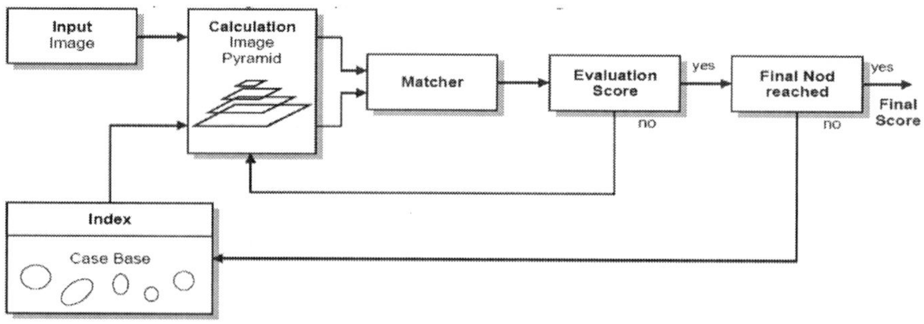

Fig. 4. Architecture of a Case-Based Object Recognition System

4.2 Image Interpretation and Results

After the methods have been recognized potential objects that are highly likely fungi spores, we are extracting more features from the objects that distinguishes the object from background and different fungi spores. Of course, one features is already the shape information used in the matching process but that is not enough for more detailed recognition. The features that we are calculating for this kind of objects is the inner structure, texture and gray level information based on our novel Random Set texture descriptor [8]. Based on this feature set we can construct the classifier. We use decision tree induction [5] based on our tool Decision Master [6]. This gives us a good classifier.

As the result we get the information about the kind of fungi spores contained in the image and the number of fungi spores versa the kind of fungi spores. The error rate was 82% by test and train [5].

5 Mapping of Image Information to Hygiene Relevant Parameter (HRP) with Data Mining

5.1 Determination of the Relationship of the Parameters to HRP with Standard Immunological Tests

Now we are mapping the outcome of the automatic system (number of fungi spores and other objects) to the hygiene-relevant parameter. The kind and the number of fungi spores was determined manually to get a true class label for each object. We want to figure out if the proposed methods can bring out information about hygiene-relevant parameters and besides that new information that can be used to control the quality of food stuff. From the 4x10 different samples were created a data base where the columns of each entry shows the class, that is the optical visual inspection label, the number of Fusarium spores, the number of Alternaria/Ulocladium, the number of Aspergillus/Penicillium, the number of Cladosporium, the number of fungi spores with unknown classification and the total number of fungi spores. Besides that, for each charge was determined the Don value based on ELISA test. In addition to the enumeration of fungal spores the concentration of a main mycotoxin of the genus Fusarium deoxynivalenol (DON) was determined. by a commercial enzyme immuno-assay screening.

Table 1-4 show that there is a significant difference in the number and the kind of fungi spores for the different charges. Figure 6 shows that DON value corresponds to the visual determined class labels. Grain with a low number of Fusarium spores have low DON values and grain charges with high number of Fusariam spores have high DON values.

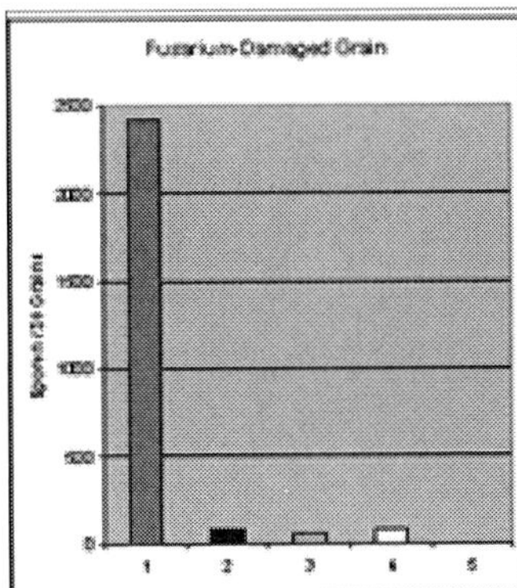

Table 1 Fusarien Damaged Graind

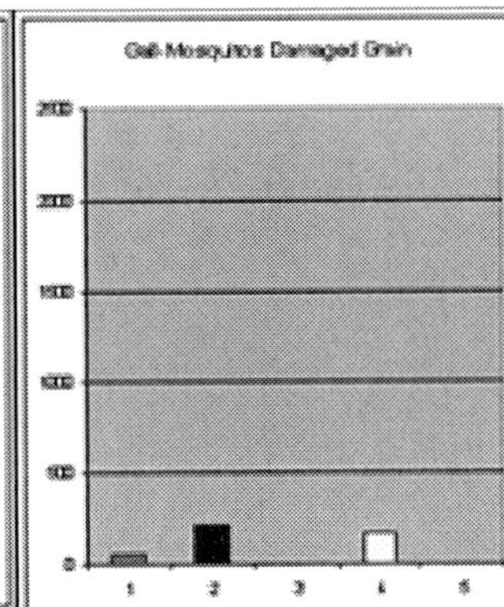

Table 2 Gall-Mosquitos Damaged Grain

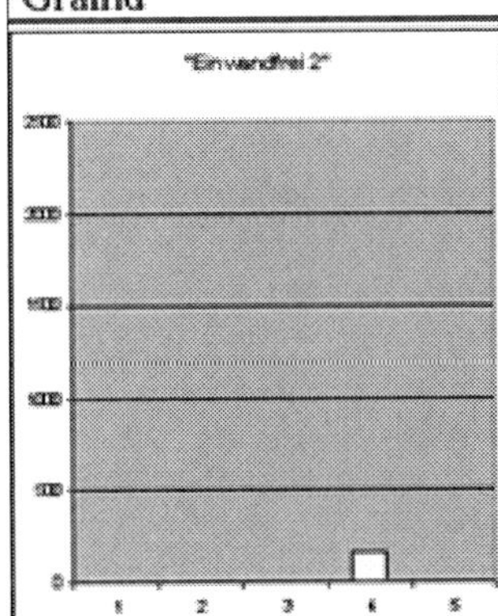

Table 3 Charge "einwand-frei 2"

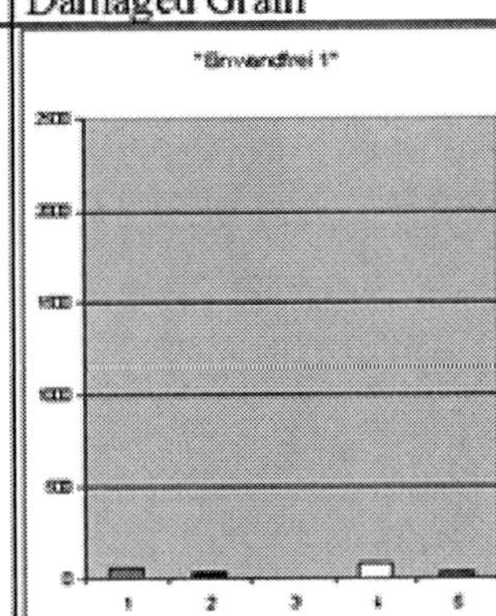

Table 4 Charge "einwand-frei 1"

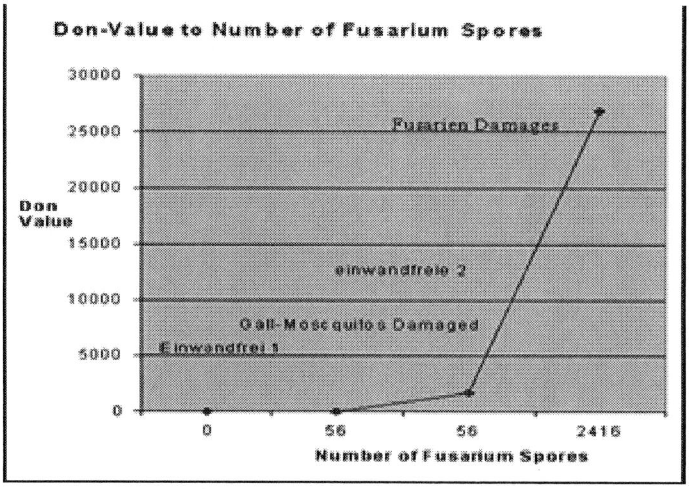

Fig. 6. Don Value to Number of Fusarium Spores

Decision tree induction [5] with Decision Master [6] on an entropy-based criterion was performed in order to find out the relation between the coverage of fungi spores and the class label (mycotoxin value).

5.2 Result of the Mapping of Image Outcome to Hygiene Relevant Parameter´s

The induction experiment shows that there is a relation between the number of Cladosporium spores and Fusarium spores respective the class, see Figure 7. It says that grain charges with a high number of Cladosporium spores will have a low number of Fusarium spores. That means these charges are either perfect charges or gall-mosquitoes damaged charges. Whereas charges with low Cladosporium spores can be either charges with a high number of Fusarium spores or a low number of Fusarium spores. Note that charge "einwandfrei 2" (visual perfect grains) has been taken out from a sample with Fusarium damaged grains. It seems that the number of Cladosporium spores indicates this fact. The number of Alternaria and Aspergillus spores did not have a significant influence in this experiment.

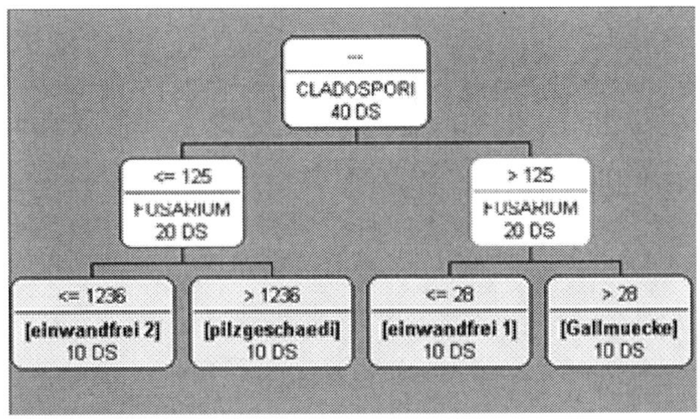

Fig. 7. Decision Tree for the Determination of Grain Quality based on Number and Type of Fusarium Spores

6 Conclusions

We have presented our results on our case study for the detection of hygiene-relevant parameters from cereal grains based on intelligent image acquisition and interpretation methods as well as data mining method. It is a joint work between computer scientist, food experts and microbiologists. We have shown that probe handling and data acquisition is an important task. The image acquisition method we have demonstrated in this paper works well and can be fully automated. It can also be constructed in such a way that the coverage from each single grain can be taken off and evaluated based on the intelligent image interpretation and data mining methods. The image analysis on case-based object recognition works well for this task but has to be tuned, so that a better object recognition rate can be achieved. From each single object can be extracted image features such as the shape and texture description. These features can be used for classification. It is preferable to construct the classifier based on decision tree induction methods. Once the type and number of fungi spores has been determined this information can be set into relation with the hygiene-relevant parameters. We have shown that the number of Fusarium spores correlate with the DON levels which is a value used for the determination of the mycotoxin concentration. However, when considering this experiment as a data mining experiment and applying decision tree induction to the created data base some other important information can be extracted which are more or less hidden before.

The aim of our work was to come up with a new measurement method for the determination of hygiene-relevant parameter´s on grains. Besides that, we like to discover formerly unknown relations or information based on the material in the coverage of the grain such as different types of fungi spores.

Acknowledgement

We like to thank the researchers from the Bundesanstalt für Bundesforschungsanstalt für Ernährung und Lebensmittel, Institut für Getreide-, Kartoffel-und Stärketechnologie in particular, Prof. Lindhauer for providing some of the examples of the material and for encouraging advices and useful discussions on the direction of the research.

References

1. Müller, H.M. et al. (1997): Fusarium toxins in wheat harvested during six years in an area of Southwest Germany. Natural Toxins 5, 25-30
2. Hermann, W.; Kübler, E. und Auf-hammer, W. (1998): Ährenbefall mit Fusarien und Toxingehalt im Korngut bei verschiedenen Wintergetreidearten. Pflanzenwirtschaften 2, 97-107
3. Rodeman, B. (2003): Auf resistente Sorten setzen. DLG Mitteilungen 3, 44-46
4. Petra Perner, Silke Jähnichen, and Horst Perner. Case-Based Object Recognition for Airborne Fungi Recognition. Intern. Journal on Artificial Intelligence in Medicine, to appear 2005
5. Petra Perner. Data Mining on Multimedia Data, Springer Verlag 2003
6. Decision Master Tool www.ibai-solutions.de
7. Petra Perner and Silke Jähnichen. Case Acquisition and Case Mining for Case-Based Object Recognition, In: Peter Funk, Pedro A. González Calero (Eds.), Advances in Case-Based Reasoning, Proceedings of the ECCBR 2004, Madrid/Spain, Springer Verlag 2004, Vol. 3155, pp. 616-629
8. P. Perner, Cognitive Aspects of Object Recognition Recognition of Objects by Texture, Eds: Liya Ding, Charles Pang, Leong Mun Kew, Lakhmi C. Jain and Robert J. Howlett, Knowledge-Based and Intelligent Information & Engineering Systems 19th Annual Conference, KES-2015, Singapore, September 2015 Proceedings, Procedia Computer Science Volume 60, p. 391-3402 (2015)

The Improved Algorithm of Barrier Tree in RNA Folding Structure Based on Basin Hopping Graph

Zhendong Liu[1,2]*, Gang Li[2], and Patrick Wang[3]

[1]School of Computer Science and Technology, Shandong Jianzhu University Jinan, 250101,China

[2]Department of Biostatistics, University of California, Los Angeles, Los Angeles, 90095,USA [3]College of Computer and Information Science, Northeastern University, Boston, 02115,USA

Email liuzd2000@126.com, patwang@ieee.org

Abstract. It is an NP-hard problem for prediction of RNA folding structure including pseudoknots, biostatistics is one method of biological data mining, the computing algorithm of RNA structure data is the important in biology. we investigate the RNA pseudoknotted structure based on characteristics of the RNA folding structure , the paper first introduce the Basin Hopping Graph(BHG) as a novel model of RNA folding landscape. Our paper gives the computing algorithm of barrier tree based on the BHG, the experimental results in Rfam13.0 and PseudoBase indicate that the algorithm is more effective. We have improved several types of pseudoknots in RNA folding structure, and analyze their possible transitions between types of pseudoknots.

Keywords: RNA Folding Structures; Pseudoknots; Algorithm; Basin Hopping Graph

1 Introduction

Basin Hopping Graph (BHG) is a novel model of RNA folding structures, computing algorithm of RNA folding structure is the important in biological data mining based on BHG. Biostatistics is one method of data mining, RNA folding is a complicated kinetic process.

Computing algorithm of RNA Structure is the important in biological data mining,biostatistics is one method of data mining.RNA folding is a complicated kinetic process. RNAs are three-dimensional molecules in biological system, which perform a wide range of function. RNA is a key component of moleculars in biological processes. The force of RNA molecules is the set of base pairs, RNA molecules can fold into a three-dimensional structure by forming base pairs of A- U,C-G match,and G-U mismatch, a pseudoknot is two overlapping base pairs, pseudoknots are pairs are known to exist in RNAs[1]. RNA secondary structures prediction is the first step to predict RNA tertiary structures in RNA sequence, RNA tertiary structures is more stable structures, some RNA folding structures are legal. It is very difficult to compute large RNA molecules including pseudoknots. It is NP-hard problem to find an optimal RNA structures. Nussinov had studied the case, where the energy function is minimized when the number of base pairs is maximized, he had designed an algorithm of $O(n^3)$ time complexity to predicting RNA secondary structures[2], Nussinov algorithm can not predict RNA structures with pseudoknots. Algebraic dynamic programming algorithm was proposed by Jens and Robert, it was used to find RNA pseudoknotted structure with simple planar pseudoknots[3], the algorithm takes $O(n^2)$ space complexity and $O(n^4)$ time complexity. The algorithm of finding optimal RNA foldings structure had been firstly known by Michael Zuker[4], Rivas and Eddy had presented Pknots algorithm for predicting RNA pseudoknotted structures based on MFE[5], which time complexity and space comlexity are $O(n^6)$ and $O(n^4)$. The predicting problem of RNA secondary structure including pseudoknots is also NP-complete[6], maximizing the number of stacking pairs allowing pseudoknots in a planar secondary structure makes it NP-hard[7], many researchers seek for approximation algorithms for NP-hard problems. In some mimic RNA structures, pseudoknots are apparently existing[8].The heuristic algorithm for finding RNA structures with pseudoknots has been presented by Ren[9]. Several publications indicates the problem of finding the optimum structure including arbitrary pseudoknots is also NP-hard[10]. People can find the more stable structure including arbitrary pseudoknots if RNA secondary structure is modelled by maximum weighted matching[11]. Some sparse-related techniques have also been applied to RNA folding structures[12-14].

Basin Hopping Graph (BHG) is a novel model of RNA folding structures. Each vertex of the Basin Hopping Graph is a local minimum, which represents the corresponding basin in the structure. Its edges connect basins when the direct transitions between them are 'energetically favorable'. Edge weights endcode the corresponding saddle heights and thus measure the difficulties of these favorable transitions. BHG can be approximated accurately and efficiently for RNA molecules well in the length range accessible

The barrier tree has two disadvantages: (i) It neglects much of the geometric information of the RNA folding structure because the neighborhood relation between basins is ignored (ii) It is high computational cost makes it unfeasible in for RNA molecules with a more base pairs in length. The BHG can overcome these shortcomings to incorporate additional information of neighborhood.

2 Model of RNA Folding Structures

2.1 Terminology

1. RNA Secondary Structure S: Let S be a set of base pairs such as $s_i.s_j$ is a base pair, base s_i or s_j $\in\{A,C,G,U\}$, $1\leq i\leq n$.
2. BHG:Basin Hopping Graph, is a novel model of RNA folding structures.
3. MFE: Minimum Free Energy

4. Stem: the RNA structure closed by base pairs (i, j) and (k, l) ∈S, and (i, j), (i+1, j-1), ..., (k, l) are base pairs, i<k<l<j.
5. Pseudoknot: if si.sj and si'.sj'∈S,i<i'<j<j',or i'< i<j'<j, then the RNA base sequence si...si'...sj sj' composes a pseudoknot.

6.K-Stacking Pairs: In the RNA secondary structure, we use $(s_i,s_{i+1},\ldots, s_{i+k};s_{j-k},\ldots,s_{j-1}, s_j)$ to describe k consecutive stacking pairs (s_i, s_j), (s_{i+1}, s_{j-1}); (s_{i+1}, s_{j-1});(s_{i+2}, s_{j-2}); ...... ; (s_{i+k-1} ,s_{j-k+1}), (s_{i+k} ,s_{j-k}).
7. Let $S=s_1s_2\ldots s_n$, $s_{ij}=s_i\ldots s_j$.

3 Predicting Algorithm of RNA Folding Structure

RNA sequence was generated with bases A, C, G, U, it is important for RNA experiments in RNA structure prediction using energy parameters [15,16,17]. We randomly selected the RNA sub-sequences in the Rfam13 and PseudoBase to compute experiments[18,19]. The algorithm can compute RNA nested structure and pseudoknotted structure in RNA sequences. Many experiments in RNA pseudoknotted structures indicated that the algorithm has better predicting accuracy averagely. The algorithm can predict more than 4100 bases of RNA sequences. We have designed effective ways to improve the prediction accuracy for long sequences.

Four experiments in family of PseudoBase can be computed less than 15 seconds with quad-core CPU and 32G memory. The experiments show that accuracy of experiments is valuable, the predicting accuracy outperforms existing algorithms, such as PKNOTS algorithm, MWM algorithm, and ILM algorithm etc[20]. Evolutionary algorithm provide a kind of important method in the RNA structure prediction[21],the structural alignment of RNA is proved to be a useful computational technique for identifying ncRNA[22,23], the efficiency of our algorithm is faster than the other related algorithms in the RNA folding structures and target structures[21,22,23].

4 The Model of BHG in RNA Folding Structures

Lemma 1. If x,y are two local energy minima of RNAfolding structures, there exits a zig-zag path connencting x and y.

Given RNA sequence S, its energy structure L is connected. (see Fig. 1).

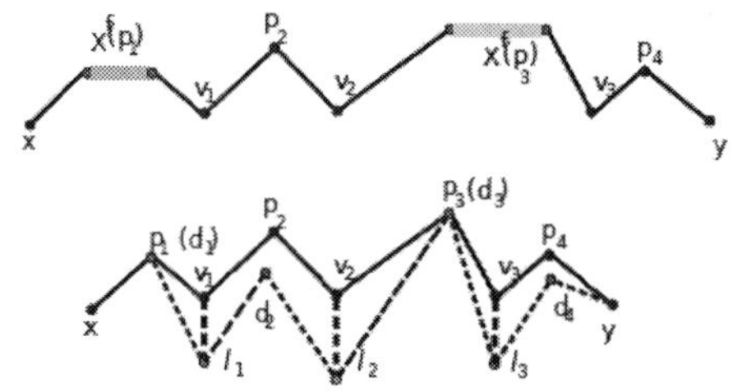

Fig. 1. the path of construction in RNA folding structure

For any substructures given from secondary structures S_1 and S_2, $S_1 \in S, S_2 \in S$, there exists a path between S_1 and S_2, for any two local energy minimum $m1 \in S1, m2 \in S2$, then there exists a zig-zag path,which connecting m_1 and m_2.

We can define the path as follow: path $P=(v_1,v_2,v_3,\ldots,v_k) \in L$, L is the RNA energy structure. If $v_i< v_{i+1}=\ldots= v_{n-1}>v_n$, then for any structures $v_{i+1}=\ldots= v_{n-1}$ are called peak points. If $v_i> v_{i+1}=\ldots= v_{n-1}<v_n$,then for any structures $v_{i+1}=\ldots= v_{n-1}$ are called valley points. If a path P fulfills three conditions:(1) max $f(vk)=S(x,y)$; (2) if $v_i< v_{i+1}=\ldots= v_{n-1}>v_n$,then each vm with $i+1\leq m\leq n-1$ is a direct saddle separating the nearest valley points that the path P passed before and after v_m; (3)if $v_i> v_{i+1}=\ldots= v_{n-1}<v_n$,then each vm with $i+1\leq m\leq n-1$, there is a minimal shelf L. we declare the path P is a zig-zag path. P can be called Basin Hopping Graph, then Basin Hopping Graph is connected .

RNA structures with pseudoknots can be generalized by the Basin Hopping Graph. We can create sets for implement the gradient walk of RNA structural class with pseudoknots, it comprises 5 types of pseudoknots as follows, Type S, Type H, Type K , Type L Type M. cf. (see Fig. 2). Type S refers to structures without pseudoknots

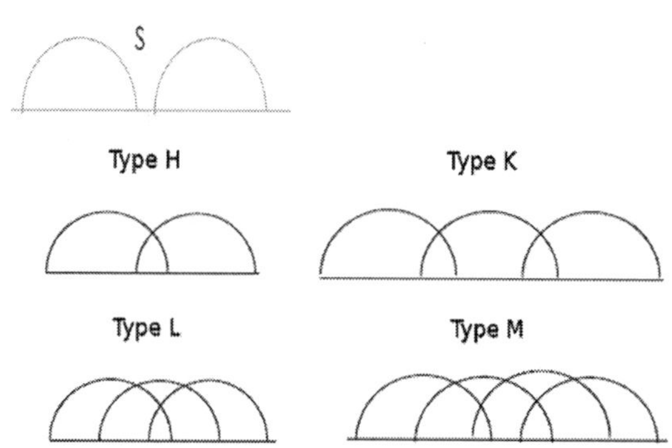

Fig. 2. types of RNA structural class

5 The Computing Algorithm of Barrier Tree Based on the BHG

Lemma 2. The barrier tree $T_b(V_b, E_b, \omega_b)$ of the RNA folding structure(X,f) is the tree Ta(Va, Ea,ωa) computed by cluster from the complete graph G(V, E,ω).

Vertex sets V_a , V_b and V were including the local minima of the RNA folding structure. Edges sets E_a ,E_b and E weight were including ω({x,y}=S(x,y) for all {x,y}∈E. S(x,y) is the saddle height between any two vertices x∈Va and y∈V_b.

Algorithm Btree T(V, E,ω)
1: M←{(x)|x∈V }
2: V_b←{ (x)|x∈M } and E_b← Ø
3: for all (x)∈V_b do
4: ω_b←f(x)
5: for all (x),(y)∈M×M do
6:$W_{x,y}$←$\omega_{x,y}$ if (x,y) ∈E and $W_{x,y}$←∞ if (x,y) E
7: while |M|>1 do
8: Search pair of clusters {(u,v)},$W_{u,v}$=min{$W_{x,y}$, (x,y) ∈C(M,2)}
9: for all (x)∈M- {(u),(v) } do
10: $W_{u,x}$= min{ $W_{u,x}$, $W_{v,x}$}
11: create new Tb-vertex(u,v) ←{u}∪ {v} with
ω b(u,v) ←$W_{u,v}$
13: V_b←V_b∪{u,v}
14: E_b←E_b∪{(u,v), (u)} ∪{(u,v), (v)}
15: M ←M-{(v)}
16:End while

The barrier tree Tb can be interpreted into a vertex weighted tree Tb-vertex with the local minima as their leaves. Internal nodes indicate the merging of BHG surrounding two local minima of saddle height. In the step of algorithm Btree, the pairs of clusters are merged which are connected by weight of the smallest edge. Weights of edge are updated to the minimum of the edge weights of the merged clusters in all vertices of separate clusters {(u,v)}. The single linkage clustering implicitly defines a binary tree T, each internal node(u,v)=(u)∪(u) is corresponding to the merging of the clusterings (u) (u), which has the minimuim weight $W_{u,v}$.

The graph B analysed by the algorithm of Btree T and obtain a binary ,vertex-weighted tree, its time complexity is $O(V^3)$.

The algorithm of Btree T has improved in time complexity and space complexity than the other barrier tree $Tb(V_b, E_b, \omega_b)$ of the RNA folding structure(X,f) .

Lemma 3. Let $BG(V_G,E_G,\omega_G)$ be the Basin Hopping Graph of the RNA folding structure(X, f), V_G denoting the sets of local minima in (X, f), then for all {x,y}∈C(V_G,2), S(x,y)=minmaxω_G({u,v}).

Lemma 4. The barrier tree Tb(Vb, Eb,ωb) of the RNA folding structure(X,f) is the tree Tc(Vc, Ec,ωc) computed by single linkage cluster from the BHG G(V, E,ω).

A Gfold software can implement Boltzmann sampling provided by Reidys. RNA folding structures can be drew by the tool VARNA. the RNA topological structures can be computed. we can study the differences in predicting RNA folding behaviour, and it can be generalized the RNA pseudoknotted framework based on BHG.

There are some examples in possible transitions between types of pseudoknots for RNAfolding strutures in real life.

Removing base pairs is relatively simple since they will never result in an invalid RNA structure, the general case involving five types of pseudoknots is rather involved, even with the restriction to RNA folding structures, with at most one pseudoknot. See Table 1.

Table 1. Possible transitions between types of pseudoknots upon removing a single base pair

Removing	M	L	K	H	S
M	1	1	1	0	0
L	0	1	0	1	0
K	0	0	1	1	1
H	0	0	0	1	1
S	0	0	0	0	1

.

Adding base pairs is also simple since they will never result in an invalid structure, the general case is five types of pseudoknots. See Table 2

Table 2. Possible transitions between types of pseudoknots upon adding a single base pair adding

Adding	M	L	K	H	S
M	1	0	0	0	0
L	1	1	0	0	0
K	1	0	1	0	0
H	0	1	1	1	1
S	0	0	1	1	1

The paper presents an exbmple named PKB92 of tobacco mild green mosaic virus, we investigate 27 bases with pseudoknots named PK1. Its RNA structure can be correctly predicted by the energy of -4.3 kcal/mol by gfold.

(see Fig. 3).

.(((((.[[[[[ )))))....]]]]].

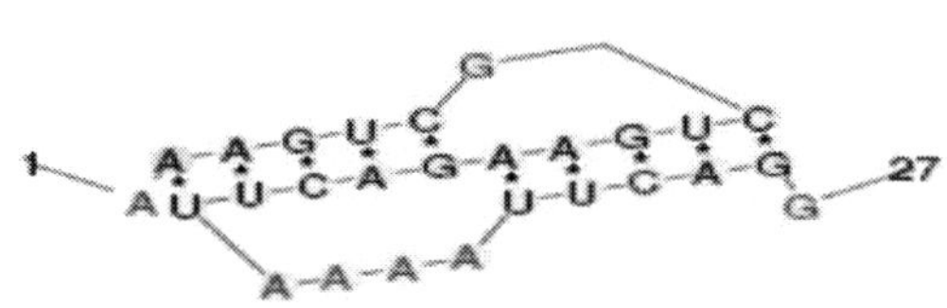

Fig. 3

The next pseudoknot-free minimum free energy, RNA secondary is with an energy of 3.9 kcal/mol.

(see Fig. 4)..
. (((((......))))) ....

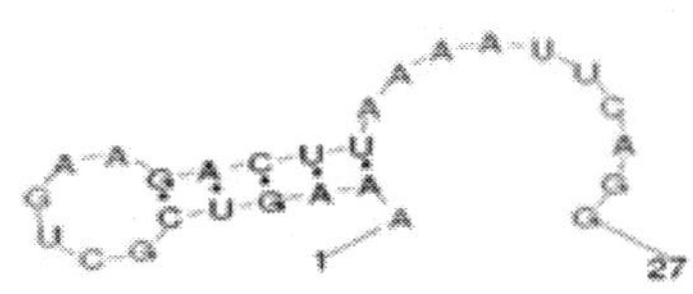

Fig.4.

It is difficulty that how to determine which base pairs can be added without changing the class of the RNA structure, and compute the changing result in energy without reevaluating the RNA folding structure. We restrict the subset of RNA structures with H-type pseudoknots in restricted class.

(A)

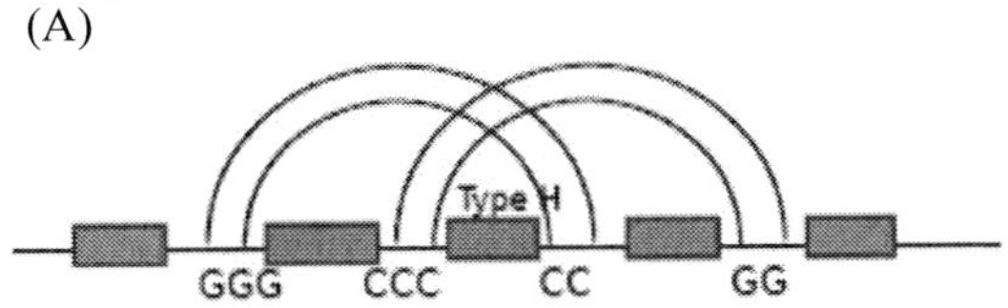

Fig. 5.

An H-type pseudoknots divide the RNA sequence into five regions: two external regions (blue) and three internal regions (green); There are two basic ways to add base pairs. (see Fig. 5).

(B)

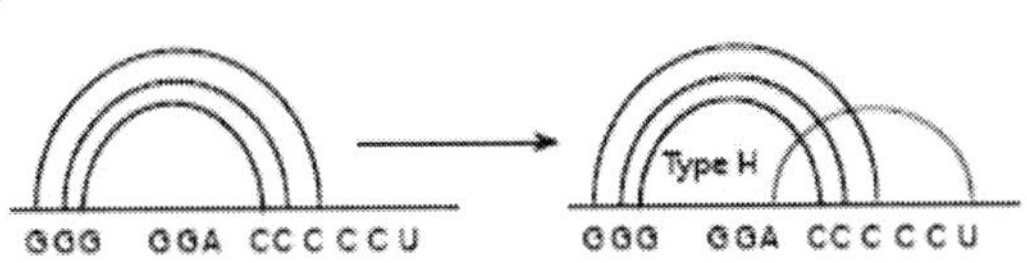

Fig. 6.

Adding a base pair crossing a stack results in an H-type Pseudoknot in RNA sequences. (see Fig. 6).

(C)

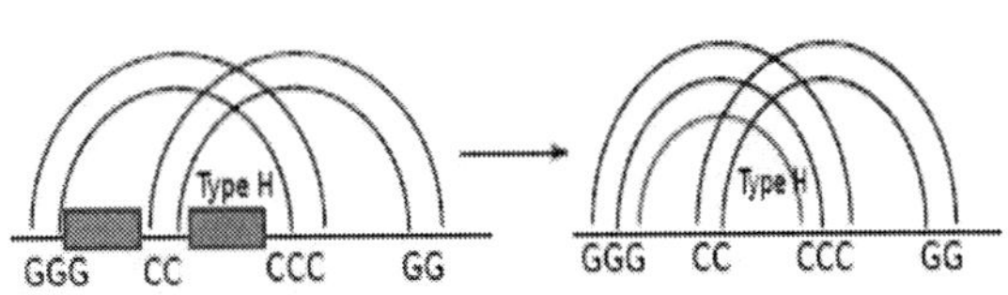

Fig. 7.

Add a base pair which involves two green regions to the existing stacks. (see Fig. 7).

(D)

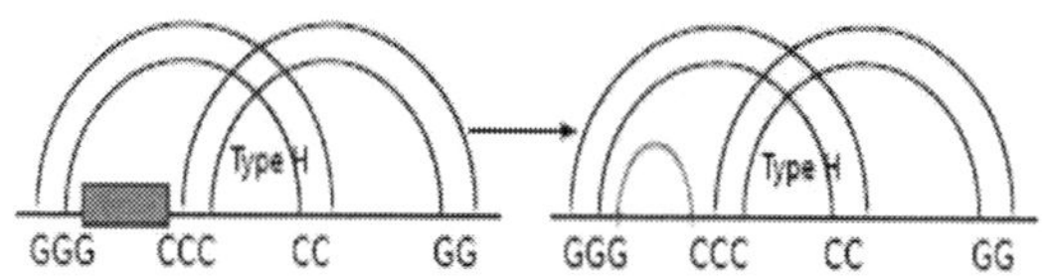

Fig. 8.

Add a base pair which involves nucleotides exactly in one green region without crossing with other existing base pairs. (see Fig. 8).

(E)

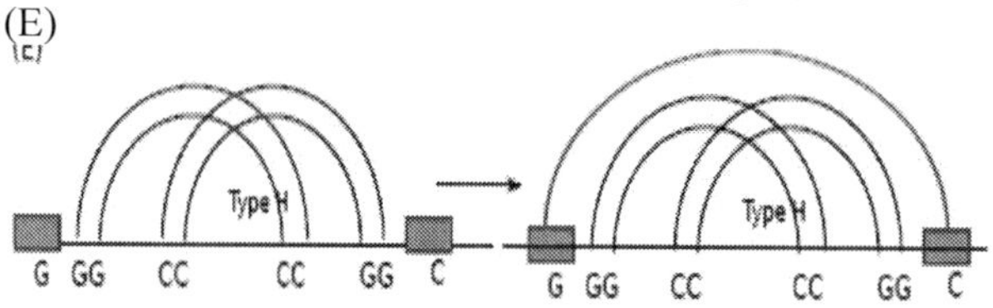

Fig. 9.

Add a base pair which involves two blue regions without crossing other existing base pairs. (see Fig. 9).

According to the principle of the BHG and MFE, the paper provides a path-searching algorithm to connect the graph LM. We investigate the low energy part of the BHG for PKB92 sequence, the PKB92 is more likely to fold the most stable secondary structure, and refold to form the pseudoknots.

6 Conclusion and Future Work

The paper has presented an efficient algorithm for predicting RNA structure with pseudoknots, the predicting accuracy, the time complexity and space complexity outperform existing algorithms, such as MWM algorithm, PKNOTS algorithm and ILM algorithm, Our paper has improved several types of pseudoknots considered in RNA folding structure, and analyze their possible transitions between types of pseudoknots, we also presented the computing algorithm of barrier tree based on the BHG.

It is a also efficient computational method for characterizing the RNA folding structure based on basin hopping graph[24]. The RNA adopts an unexpected tandem three-way junction RNA structure, and unspliced dimeric genomes selected by the RNA conformer may direct packaging[25].

Given an underlying model of gene expression, BayFish uses a Monte Carlo method to estimate the Bayesian posterior probability of the model parameters in smFISH data of single-molecule RNA, RNA Sequencing Reveals and RNA Polymerization in tRNA Fidelity and Repair also are important for RNA structure prediction [26,27].

In the future, we should improve predicting accuracy of the algorithm of RNA folding structures with pseudoknots, and improve the computing algorithm of barrier tree based on the BHG. We should also improve Monte Carlo method to estimate the Bayesian probability of the model parameters in smFISH data of single-molecule RNA in single cells. We also focus on the RNA Sequencing Reveals and their applca-tion.

7 Acknowledgements

Our work was supported by the National Natural Science Foundation of China under Grant No.61672328,61672323. We are grateful to T.Akutsu, Lyngsø, Ieong Rivals, and Kucharík for their efficient work.

References

1. R.Nussinov, G.Pieczenik, J.Griggs and D.J.Kleitman.: Algorithms for loop matchings, SIAM J. Appl. Math 35(1), 68- 82 (1978).
2. Michael Zuker.: On finding all suboptimal foldings of an RNA molecule. Science 244, 48-52(1989).
3. E.Rivas and S.R.Eddy.:A Dynamic Programming Algorithm for RNA Structure Prediction Including Pseudoknots. Journal of Molecular Biology 285, 2053-2068(1999).
4. Lyngsø and N. S Christian.: Pseudoknots in RNA Pseudoknotted Structure, Proceedings of Recomb. Tokyo Publishing 201-209(2000).
5. S.Ieong, M.Y.Kao, T. W.Lam, et al.: Prediction RNA pseudoknotted structures with arbitrary pseudoknots by maximizing the number of stacking pairs. Journal of Computational Biology 6, 981-995(2003).
6. M. H.Kolk, M.vanderGraff, S.Wijmenga, et al.: NMR structure of a classical pseudoknots interplay of single and double-stranded RNA. Science 280, 434-438(1998).
7. Jihong, Ren, Baharak Rastegari, Anne Condon et al.:HotKnots:Heuristic prediction of RNA pseudoknotted structures including pseudoknots. RNA, 1494-1504(2005).
8. T.Akutsu.: A dynamic programming algorithm for RNA structure prediction with pseudoknots. Discrete Applied Mathematics 104, 45-62(2000).
9. J.E.Tabaska, R.B.Carry, H.N.Gabow and G.D.Stormo.: An RNA folding method capable of identifying pseudoknots and base triples. Bioinformatics 14,691-699(1998).
10. Y.Zhang, Y.M.Cheung, B. Xu and W.F. Su.:Detection Copy Number Variants from NGS with Sparse and Smooth Constraints. IEEE/ACM Transactions on Computational Biology and Bioinformatics 14(4), 856-867(2017).
11. R.Backofen, S. D.Tsur, S.Zakov and M.Ziv-Ukelso.: Sparse RNA Folding: Time and Space Efficient Algorithms, Annual Symposium on Combinatorial Pattern Matching, 249-262(2009).
12. R.Backofen, S. D.Tsur, S. Zakov and M.Ziv-Ukelson .: parse RNA Folding: Time and Space Efficient Algorithms. Journal of Discrete Algorithms 9(1), 12-31(2011).
13. D.H.Turner, N.Sugimoto and S.M.Freier.: RNA Structure Prediction. Annual Rewiew of Biophysics Chemistry 17, 167-192(1998).
14. J.A.Jaeger, D. H.Turner and Zuker.: Improved predictions of pseudoknotted structures for RNA. Proc Natl Acad Sci 86, 7706-7710(1989).

15. J.Ruan, G. D.Stormo and W. Zhang.: An Iteratted loop matching approach to the prediction of RNA Pseudoknotted Structures with pseudoknots. Bioinformatics 20, 58-66 (2004).
16. http://www.bio.leidenuniv.nl/~Batenburg/PKBGet.html.
17. Zhendong Liu, Hengwu Li and Damig Zhu.: A Predicting Algorithm of RNA Pseudoknotted Structure Based on Stems. Kybernetes 39(6), 1050-1057(2010).
18. B.Han.: Structural alignment of pseudoknotted RNA. J. Comput. Biol 15, 489-500(2008).
19. Yuping Wang and Chuangyin Dang.:An Evolutionary Algorithm for Global Optimization Based on Level-Set Evolution and Latin Squares. IEEE Transactions on Evolutionary Computation 11(5), 579-595(2007).
20. T.K.Wong.: Structural alignment of RNA with complex pseudoknot structure. J. Comput. Biol. 18, pp.97-108,(2011).
21. Zhendong Liu,Daming Zhu,Hongwei Ma.: Predicting Scheme of RNA folding Structure including Pseudoknots. International Journal of Sensor Networks 16(4), 229-235(2014).
22. Marcel Kucharik, Ivo L. Hofacker,Peter F. Stadler and Jing Qin.:Basin Hopping Graph: A computational framework to Characterize RNA folding landscapes. Bioinformatics 30(14)2009-2017 (2014).
23. C. Sarah, Keane, Xiao Heng, Kun Lu. et al.: Structure of the HIV-1 RNA packaging Signal. Science 348(6237), 917-921(2015).
24. Mariana Gómez-Schiavon, Liang-Fu Chen, Anne E. West and Nicolas E. Buchler.: BayFish: Bayesian inference of transcription dynamics from population snapshots of single-molecule RNA FISH in single cells. Genome Biology 18:164 (12 pages) (2017)
25. Zhendong Liu,Daming Zhu and Qionghai Dai.: Predicting Model and Algorithm in RNA Folding Structure Including Pseudoknots. International Journal of Pattern Recognition and Artificial Intelligence 32(10), (17 pages) (2018).
26. Yury V. Malovichko, Kirill S. Antonets, Anna R. Maslova, Elena A. Andreeva, Sergey G. Inge-Vechtomov and Anton A. Nizhnikov.: RNA Sequencing Reveals Specific Transcriptomic Signatures Distinguishing Effects of the [SWI+] Prion and SWI1 Deletion in Yeast Saccharomyces cerevisiae. Genes 10(3), 212(2019).
27. Allan W. Chen, Malithi I. Jayasinghe, Christina Z. Chung, Bhalchandra S. Rao, Rosan Kenana, Ilka U. Heinemann and Jane E. Jackman.: the Role of 3′ to 5′ Reverse RNA Polymerization in tRNA Fidelity and Repair. Genes 10(3), 250(2019).

Automatic Liver Segmentation Using shape context constraint Network*

Lifang Zhou[1,2,3], Lu Wang[1], Patrick Wang[4], Weisheng Li[1], Bangjun Lei[2], and Xueyuan Deng[1]

[1] Coll. of Computer Science and Technology, Chongqing Univ. of Posts and Telecommunications, Chongqing 400065, China
zhoulf@cqupt.edu.cn

[2] Hubei Key Laboratory of Intelligent Vision Based Monitoring for Hydroelectric Engineering, China Three Gorges University, Yichang 443002, China

[3] Coll. of Software, Chongqing Univ. of Posts and Telecommunications, Chongqing 400065, China

[4] Univ. of Northeastern university, Boston, MA 02115, USA

Abstract. Neural Networks, as one of the powerful tool in medical images segmentation, has received considerable attention. While in real scenarios, the amount of label data produced by physicians is limited. To alleviate this problem, we propose a shape context-assisted neural network (SNN), which is consisted of a single-slice segmentation predication model and a context-shape predication model. First, the middle slices is focused by the network so that the adjacent slices of them will be constrained. Next, the training data of middle slices are used to obtain a prediction model by the strategy of expanding and dropout. Finally, the context-shape prediction model is constructed based on a network architecture of cyclic multiple input so that the shape constraint among consecutive slices can be achieved. The results indicate that our proposed method outperforms not only the traditional 2D-based network significantly but also the 3D-based network slightly in the situation of small samples.

Keywords: Liver segmentation · Context shape constraints · Neural network.

1 Introduction and Related Work

Automatic liver segmentation from computed tomography (CT) images is essential to various medical procedures, such as diagnosis, treatment planning, and image-guided surgery. However, the segmentation of liver is still a challenge task because of three factors, such as low intensity contrast between liver and other neighboring organs, large anatomical variation in both shape and size, and

* Supported by Chongqing Overseas Scholars Innovation Program(cx2018124), Hubei Key Laboratory of Intelligent Vision Based Monitoring for Hydroelectric Engineering(2017SDSJ02) and Chongqing research and innovation project of graduate students(CYS18247).

the presence of noise [15]. Building a robust system that overcomes the above difficulties is still an open problem.

Recently, full convolutional neural networks (FCNs) have shown obvious advantage on a range of recognition problem [8]. Inspired by this, many researchers have made effort to segment liver and tumor by using deep learning methods. For example, Sun et al.[13] designed a multi-channel FCN to segment liver tumors from CT images, in which the probability map was generated by features fusion from multiple channels. It can learn the unique information of the pathological features from contrast-enhanced data and improve the segmentation accuracy. It is well known that the connection between the upper and lower slices is critical for liver segmentation due to the importance of spatial information. The information along the z-axis should be considered carefully like radiologist. Prasoon et al.[1] applied three 2D FCNs on orthogonal planes (e.g., the xy, yz, and xz planes), so the voxel prediction results will be generated by the average of these probabilities. The method utilizes spatial shape information of 3D data by segmenting 2D slices of different planes. Besides, Milletari et al [9] proposed a localization and segmentation method from a CT volume utilizing 3D FCN features for localization. Compared to 2D based methods [3, 2], the depths of network as well as filter's field of view have been limited because of high computational cost and GPU memory consumption from 3D-based methods [6, 11]. Unfortunately, they are essential factors to the performance of segmentation.

In this paper, we propose a novel cascaded network, called shape context-assisted neural network (SNN), where intra-slice features and 3D contexts are effectively probed and jointly optimized for accurate liver segmentation. The network contains two models, a single-slice model and a context-shape model. The core of our network for liver segmentation is to combine the advantage of segmentation network with a context-shape prediction network to attain the coherence information. Experimental results demonstrate that the SSN model outperforms the state-of-the-art methods on the 3Dircadb and LiTs database. In addition, it is significant benefit for the flexibility of context-shape to integrate into any segmentation network.

2 Method

Our pipeline is illustrated in Figure 1, which is consisted of the single-slice model and context-shape model. In single-slice stage, the coarse segmentation of liver can be attained by training the middle slice of a CT volume. Next, the context-shape model will be proposed to probe intra-slice and inter-slice features through shape constraint loss and pixel-wise loss. In fact, we adopt a cascaded learning strategy, so that the slices of different liver region can be segmented through a coarse-to-fine manner.

2.1 Network architecture

The single-slice model and the context-shape model are both based on U-net[10], which is an architecture for cell image segmentation. In our network, since the

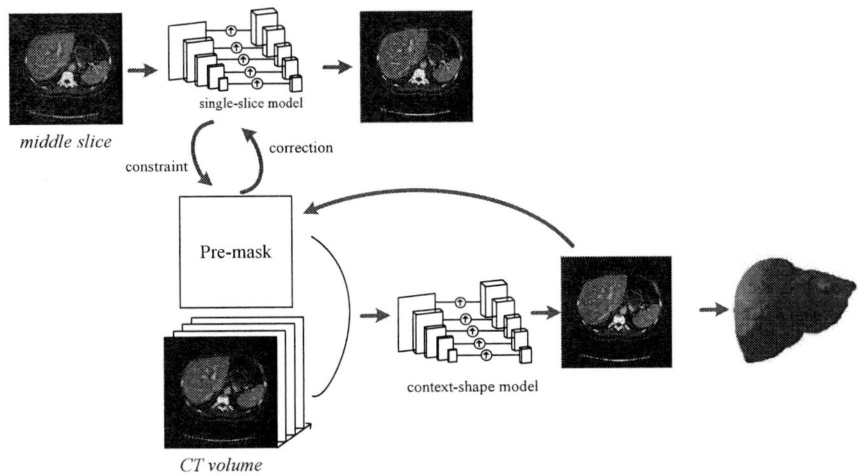

Fig. 1. The illustration of the proposed network architecture.

network structure of the first half of U-net is same as that of the first ten layers of VGG16 [12], so we have replaced it. The weight of VGG16 pre-training model can be loaded for initialization, so the weights only need to be fine-tuned during training. It can reduce the disturb because of the number of training samples on the model.

More details are described as follows. Single-slice model only trains and predicts the slice of the intermediate sequence including the large liver region. The context-shape model continuously trains and predicts the slice of the non-intermediate sequence. Assuming that the number of slices containing liver in i^{th} CT data is N_i, $i = 1, 2, \cdots, n$, and the number of the middle slice is $N_i/2$. Here, we set it to p_i, $i = 1, 2, \cdots, n$, and $p_i \in P$. In order to achieve higher segmentation accuracy with fewer samples, the data P is expanded to P'. Meanwhile, the dropout idea [5] is employed to prevent over-fitting. Some slices in P' are discarded based on a certain ratio.

In the training stage, three parts are used to train the context-shape model. They are the original slice of p_i, its groundtruth, and the groundtruth of slice $p_{i\pm1}$ respectively. The loss function is defined as the loss of two parts: the original slice with groundtruth, and the shape constraint with groundtruth. As a result, the correlation between adjacent slice and current slice have been trained.

2.2 Loss object

The input images and their corresponding segmentation maps are used to train the network with the stochastic gradient descent implementation of Keras. The energy function is computed by a pixel-wise soft-max combined with the cross entropy loss function. The soft-max is defined as

$$a_i = \frac{e^{z_i}}{\sum_k^K e^{z_k}} \tag{1}$$

where z_i denotes the activation in feature channel i, and a_i is the approximated maximum-function. $a_i \approx 1$ for the i that has the maximum activation z_i and $a_i \approx 0$ for all other i.

$$C = -\sum_i y_i \ln a_i \tag{2}$$

where y_i is the true label of each pixel and a_i is a weight map.

In context-shape model, there are shape constraints. So, the loss function can be defined by

$$C = -\sum_i y_i \ln a_i - \lambda \sum_i \hat{y}_i \ln a_i \tag{3}$$

where y_i is the true label of each pixel, a_i is a weight map and $\hat{y}_i$ is the shape constraint, which means the pre-mask input. λ is a number between 0 and 1, which balances the weight of shape constraints in the segmentation process. Generally, as λ increases, shape constraint of adjacent slices has a greater influence on the segmentation result, and its spatial constraint becomes stronger. We set up experiments to determine the value of the weight. It should be noticed that shape constraint is idea when weight has reached 0.3. Notably, loss function can be solved by the following formula.

$$\frac{\partial C}{\partial z_i} = \sum_j \left(\frac{\partial C_j}{\partial a_j} \frac{\partial a_j}{\partial z_i} \right) = a_i \sum_i (y_j + \lambda \hat{y}_j) - y_i - \lambda \hat{y}_i \tag{4}$$

2.3 Data Preprocessing

Data augmentation is essential for the network to attain the properties of invariance and robustness because the number of training samples is always shortage in medical image segmentation. Therefore, we proposed to augment the slices with large liver region, which can improve the robustness of the model for shift, rotation invariance, elastic deformations, and gray value variations. In detail, we have built a three-channel images. The first two channels both are the original slices and the other is the corresponding groundtruth. Samples can be obtained by warping, rotating, translation, shearing, and scaling. Furthermore, the correspondence between the original slice and ground truth can be guaranteed. However, the main limitation of augmented slices is tendency to attain a single shape. Next, in order to get the compact data, a big data dropout strategy (BDD) [7] is introduced in multi-data images.The pipeline is illustrated in Figure 2.

Fig. 2. The illustration of Data Preprocessing.

3 Results

3.1 Effectiveness of SNN

Regarding the performance of images form 3Dircadb dataset, we first verified the effectiveness of a single-slice model, as we observed in Figure 3. The data in the first and second rows are selected from two random CT volumes in 3Dircadbs database. The red curve and green curve denote the groundtruth and the segmentation result, respectively. When the red line coincides with the green line, only the green line is shown, indicating the best segmentation result. Figure 3 shows 4 columns that demonstrate the segmentation results by training original middle slices, augmented slices, augmented slices with BBD and the correction of 3(c), respectively. Furthermore, the superiority of the proposed method can be found by each of tested steps respectively. Compared with 3(a), the segmentation accuracy of 3(b) is improved sharply, which shows the importance of utilizing the augmented strategy. From Figure 3, it is clearly that the segmentation result is further improved by BDD strategy. Finally, the test result of 3(d) demonstrate that the context segmentation model can contribute to help the network achieve the promising result.

3.2 Quantitative results and comparisons

Evaluation metrics In order to more easily emphasize advantages and disadvantages of the proposed method, five different error measures have been used to measure the volumetric overlap or surface distances of the segmentation results(A) as compared to the ground truth(B)[14]. Among them, Overlap Error (VOE) measures the percentage of all the number of overlapping voxels in A and B, where 0 means perfect results. Relative Volume Difference (VD) measures the relative volume difference between A and B, which can determine the severity of over-segmentation ($VD > 0$) and under-segmentation ($VD < 0$). Average Symmetric Surface Distance (AvgD) is used to measure the sum of the shortest surface distances between A and B. The more similar the segmentation results are, the smaller value have. Maximum Symmetric Surface Distance (MaxD) can determine the maximum surface distance between A and B, which is sensitive to the abnormal segmentation results. Dice Score (DICE) is a statistic used to gauge the similarity of A and B, where 1 means the perfect segmentation. The

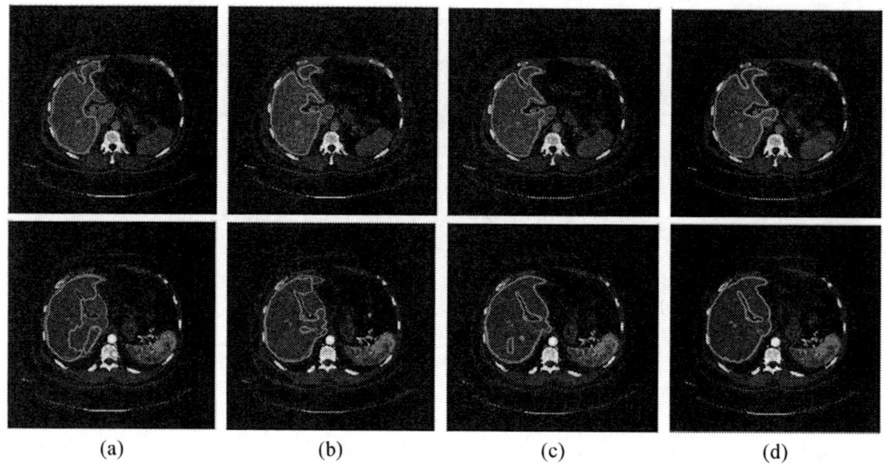

Fig. 3. The illustration of single-slice model with different training data.

Jaccard similarity coefficient(Jaccard) is used to compare similarities and differences between A and B. The larger Jaccard coefficient value have, the higher similarity of samples are.

Comparison in 3Dircadb database To validate the effectiveness of the small sample, we compare the segmentation result of FCN and U-net. In detail, the training data have included only 16 CT volumes of 3Dircadb database, and the remaining 4 CT volumes are used to test. As illustrated in Table 1, five evaluation of FCN and U-net are all relatively poor than ours. In detail, it is because the spatial continuity between different slices A and B have absent. Besides, U-net has serious over-segmentation and FCN exists under-segmentation, which are all related to the small number of training samples. In fact, the superiority of ours resulted from the fact that the large variation of liver region can be avoided. It is because that slices including either larger liver region or smaller liver region have been trained and segmented separately. In addition, the correction of the context-shape model for integration into middle slices segment results is a significant benefit.

Table 1. Compared with FCN and U-net in 3Dircadb.

methods	VOE	VD	AvgD	MaxD	Dice	Jaccard
FCN	13.78 ± 7.2	-10.89 ± 5.8	3.42±1.4	64.81±22.7	83.26±7.2	75.85± 8.5
U-net	39 ± 9.2	87± 42.5	19.4±12.3	119 ± 67.7	72.9 ± 12.9	68.98±7.3
Ours	6.23 ± 2.3	0.9± 3.1	1.3 ± 0.6	24.3± 8.0	95.9± 2.2	92.53 ± 2.9

Comparison in LiTs database We have also validated the proposed method on LiTs database by randomly selecting 80 volumes for training and 20 volumes for testing. Table 2 shows the mean quantitative comparative results of 3D-unet[4], V-net and ours. Compared with 3D-unet and V-net, the proposed method is only slightly inferior to the evaluation of AvgD, while the other four assessment indicators are better. Since 3D-based method considers spatial coherence completely, 3D-unet and V-net methods have a dice score higher than 90. Nevertheless, the dice score of the proposed method, aiming at small sample set, is slightly improved by training 70 sets. It can be seen that the proposed method are not inferior to 3D-unet and V-net through increasing contextual shape correlation.

Table 2. Compared with 3D-unet and V-net in LiTs.

methods	VOE	VD	AvgD	MaxD	Dice
3D-unet	11.44± 4.73	2.25±3.67	1.9±1.93	33.06±8.77	92.1±3.9
Vnet	9.27±1.88	1.57±2.81	1.4±0.63	24.55±4.21	95.12±1.91
Ours	7.89±1.52	1.1±2.43	1.5±0.5	23.21±4.73	96.5±1.95

4 Conclusions

In this work, we have proposed an algorithm to segment liver in 2D slices using two cascaded segmentation networks. The network makes full use of the spatial coherence of the adjacent slice, which can improve the accuracy of liver segmentation. Compared with traditional 2D-based network, the segmentation accuracy is greatly improved in the situation of small sample set. Compared with 3D-based network in the situation of large sample set, the proposed method showed higher dice coefficients than others again.

Although encouraging results have bee achieved, the accuracy of segmentation still needs to be further improved(see Table 1 and Table 2). Furthermore, it is important to take into account data diversity. Therefore, augmenting data using artificial multi-modal, extracting different modal features and exploring their contribution for the accuracy of liver segmentation are all essential for our future work.

References

1. Adhish, P., Kersten, P., Christian, I., Fran?Ois, L., Erik, D., Mads, N.: Deep feature learning for knee cartilage segmentation using a triplanar convolutional neural network. In: Medical Image Computing & Computer-assisted Intervention: Miccai International Conference on Medical Image Computing & Computer-assisted Intervention (2013)
2. Ben-Cohen, A., Diamant, I., Klang, E., Amitai, M., Greenspan, H.: Fully convolutional network for liver segmentation and lesions detection. In: International Workshop on Large-scale Annotation of Biomedical Data & Expert Label Synthesis (2016)

3. Christ, P.F., Elshaer, M.E.A., Ettlinger, F., Tatavarty, S., Bickel, M., Bilic, P., Rempfler, M., Armbruster, M., Hofmann, F., DAnastasi, M.: Automatic liver and lesion segmentation in ct using cascaded fully convolutional neural networks and 3d conditional random fields. In: International Conference on Medical Image Computing & Computer-assisted Intervention (2016)
4. Cicek, O., Abdulkadir, A., Lienkamp, S.S., Brox, T., Ronneberger, O.: 3d u-net: Learning dense volumetric segmentation from sparse annotation (2016)
5. Hinton, G.E., Srivastava, N., Krizhevsky, A., Sutskever, I., Salakhutdinov, R.R.: Improving neural networks by preventing co-adaptation of feature detectors. Computer Science **3**(4), pgs. 212–223 (2012)
6. Hu, P., Wu, F., Peng, J., Liang, P., Kong, D.: Automatic 3d liver segmentation based on deep learning and globally optimized surface evolution. Physics in Medicine & Biology **61**(24), 8676 (2016)
7. Li, X., Dou, Q., Chen, H., Fu, C.W., Qi, X., Belavy, D.L., Armbrecht, G., Felsenberg, D., Zheng, G., Heng, P.A.: 3d multi-scale fcn with random modality voxel dropout learning for intervertebral disc localization and segmentation from multi-modality mr images. Medical Image Analysis **45**, 41 (2018)
8. Long, J., Shelhamer, E., Darrell, T.: Fully convolutional networks for semantic segmentation. IEEE Transactions on Pattern Analysis & Machine Intelligence **39**(4), 640–651 (2014)
9. Milletari, F., Navab, N., Ahmadi, S.A.: V-net: Fully convolutional neural networks for volumetric medical image segmentation. In: Fourth International Conference on 3d Vision (2016)
10. Ronneberger, O., Fischer, P., Brox, T.: U-net: Convolutional networks for biomedical image segmentation (2015)
11. Roth, H., Oda, M., Shimizu, N., Oda, H., Hayashi, Y., Kitasaka, T., Fujiwara, M., Misawa, K., Mori, K.: Towards dense volumetric pancreas segmentation in ct using 3d fully convolutional networks. In: Medical Imaging 2018: Image Processing. vol. 10574, p. 105740B. International Society for Optics and Photonics (2018)
12. Simonyan, K., Zisserman, A.: Very deep convolutional networks for large-scale image recognition. Computer Science (2014)
13. Sun, C., Guo, S., Zhang, H., Li, J., Chen, M., Ma, S., Jin, L., Liu, X., Li, X., Qian, X.: Automatic segmentation of liver tumors from multiphase contrast-enhanced ct images based on fcns. Artificial Intelligence in Medicine **83**, S0933365716305930 (2017)
14. Tobias, H., Bram, V.G., Styner, M.A., Yulia, A., Volker, A., Christian, B., Andreas, B., Christoph, B., Reinhard, B., Gy?Rgy, B.: Comparison and evaluation of methods for liver segmentation from ct datasets. IEEE Transactions on Medical Imaging **28**(8), 1251–1265 (2009)
15. Zhou, L., Qi, Z., Li, W.: Automatic segmentation for medical image with the optimized tree structured part model. In: International Conference on Biomedical Engineering & Informatics (2016)

Author´s Index

Bilbao	Javier	10
Bilbao	Imanol	10
Ciufudean	Calin	1
Deng	Xueyuan	41
Feniser	Cristina	10
Lei	Bangjun	41
Li	Gang	31
Li	Weisheng	41
Liu	Zhendong	31
Olozaga	Andoni	10
Perner	Petra	23
Wang	Patrick	31, 41
Wang	Lu	41
Zhou	Lifang	41

Journals by ibai-publishing

The journals are free on-line journals but having in parallel hardcopies of the journals. The free on-line access to the content of the paper should ensure fast and easy access to new research developments for researchers all over the world. The hardcopy of the journal can be purchased by individuals, companies, and libraries.

Transactions on Machine Learning and Data Mining
(ISSN: 1865-6781)

The International Journal "Transactions on Machine Learning and Data Mining" is a periodical appearing twice a year. The journal focuses on novel theoretical work for particular topics in Data Mining and applications on Data Mining.

Net Price (per issue): EURO 100
Germany (per issue): EURO 107 (incl. 7% VAT)

Submission for the journal should be send to:
info@ibai-publishing.org

For more information visited: www.ibai-publishing.org/journal/mldm/about.html

Transactions on Case-Based Reasoning
(ISSN:1867-366X)

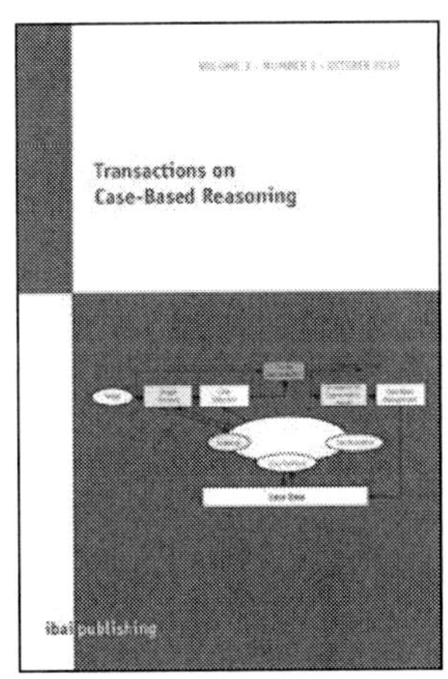

The International Journal "Transactions on Case-Based Reasoning" is a periodical appearing once a year.

Net Price (per issue): EURO 100
Germany (per issue): EURO 107 (incl. 7% VAT)

Submission for the journal should be send to:
info@ibai-publishing.org

For more information visited: www.ibai-publishing.org/journal/cbr/about.html

Announcement

World Congress DSA 2020

The Frontiers in Intelligent Data and Signal Analysis
July 12 - 23, 2020, New York, USA

www.worldcongressdsa.com

We are inviting you to our fourth World congress on the Frontiers of Signal and Image Analysis DSA 2020 to New York, Germany.

This congress will feature three events:

- the 16th International Conference on Machine Learning and Data Mining MLDM (www.mldm.de),
- the 20th Industrial Conference on Data Mining ICDM (www.data-mining-forum.de),
- and the 15th International Conference on Mass Data Analysis of Signals and Images in Artificial Intelligence&Pattern Recognition MDA-AI&PR (www.mda-signals.de).

Workshops and Tutorial will also be given.

Come to join us to the most exciting event on Intelligent Data and Signal Analysis.

Sincerely your,
Prof. Dr. Petra Perner

MLDM icdm mda

www.mldm.de www.data-mining-forum.de www.mda-signals.de